Preacher ain't done 'til the snakes are back in the box

Cover photo:

Front, Photos published with approval from News and Observer Collection, Verlin Short and Gregory Coots.

Back, Phil Dunning

MIX
Papper från ansvarsfulla källor
Paper from responsible sources
FSC® C105338

This book sprung from my genuine interest in snakes and my curiosity about history and religion. My idea with this book is to tell an interesting story, share some facts about the venomous snakes used in the Pentecostal church, and let the believers have their say. I have also included a few critical voices, and my own opinion can be found here and there.
I am not the right person to write anything close to an academic thesis, you should see this book as educational entertainment. If you are looking for statistics, tables and graphs, this is the wrong book for you.

Many thanks to everyone who helped me get the book from an idea to a finished product!
Thank you all for letting me use your photos! Also, thanks to those of you who uploaded images, free to use, online! All photographers are mentioned under their photos.

Thanks to Scott Eipper and Tie Eipper for yet another excellent proofreading and thanks to Phil Dunning for proofreading the chapters about the three species which is most common in Appalachia's snake-handling Pentecostal churches. English is not my first language, writing non-fiction books in languages other than Swedish is not easy for me. Your help was so valuable! I'm sure I've missed some corrections, you as a reader will just have to put up with the flaws in my English. And so that no shadow falls on the proofreaders, I should also add that I added some text after the proofreading.

Without an understanding family, this book would certainly not have come about, thank you Jenny, Ludwig and Jacob! I spend far too much time on herpetology, but you let me.

Finally, a big thanks to those of you who choose to read my book!
I hope you find the topic as interesting as I do! For me, the topic was so captivating that it was difficult to stop writing about it.

Best regards

Rickard, February 10th, 2025.

Rickard Ljunggren

PREACHER AIN'T DONE 'TIL THE SNAKES ARE BACK IN THE BOX

The history, the venomous snakes and the pastors' own stories

"A lot of times people get bit because they are ahead of God. They have moved on God when it was not time."

"Driving to church is pretty dangerous. But nobody's going to ban cars."

"People say we murder ourselves, but we martyr ourselves."

"The feeling to take up serpents is unexplainable, it's... it's a... peace that's surpasses all understanding, to know that you're set, that you're standing there, right there, with... with death in your hand. And the anointment of God is protected you, to let you do that."

"Our message is not 'handle snakes, handle snakes, handle snakes'. But our message is, 'Be saved by the blood of Christ.' We're not a cult. We're not freaks. We're Christians."

"That rattlesnake gave him a hard lick. In thirty seconds, he was down."

"Sometimes the Lord lets the snake bite to prove to the unbeliever that the danger's real."

Imprint

Photo: Photographers mentioned under each picture in the book.

Proofreading: Scott Eipper and Tie Eipper (whole book), Phil Dunning (chapter 8-10)

Publisher: BoD · Books on Demand, Östermalmstorg 1,
114 42 Stockholm, bod@bod.se
Print: Libri Plureos GmbH, Friedensallee 273, 22763 Hamburg, Tyskland

ISBN: 978-91-8080-889-7

"It was the most pain I guess I've ever felt in my life. During the time that it was rotting, and I knew something was going on, I just didn't know what for the first month. That much of the bone was exposed before it broke off.
My wife told me, when this broke off in the yard, she said
-I want to keep this.
I said -Why?
She said -I'll always have a piece of you no matter where you go."

Pastor Jamie Coots in an interview on ABC News. He talked about the aftermath of a venomous snake bite. At the end of his story, he held up what once was the tip of his right middle finger.

"This is my God-given right in the United States," he said on the courthouse steps before the hearing. *"If God moves on me to take up a serpent, I take up a serpent."*

Table of Contents

In a serpent handling church, one may observe not only preaching, praying aloud together on one's knees, dancing and singing but also serpent handling, fire dancing, speaking in tongues, the laying on of hands for healing, testimonies, and at times, poison drinking and foot washing. An observer may also witness stigmata, the oozing of blood from the hands or feet in empathy with the death of Jesus.

*

All quotations in the book are quoted exactly as they were written, I have not changed the choice of words or the spelling, the language is the quoted person's own.

1. Preamble

Mark 16:17-18, the King James version:

And these signs shall follow them that believe; In my name shall they cast out devils; they shall speak with new tongues; They shall take up serpents; and if they drink any deadly thing, it shall not hurt them; they shall lay hands on the sick, and they shall recover.

Photo: cgrape, Pixabay.

Verses 9-20 of chapter 16, which deal with the resurrection of Jesus, are missing from the oldest manuscripts (handwritten Bibles, before the art of printing came along). The language and style of this part of the Gospel of Mark differs from the language and style of the rest of the Gospel, and the verses form a summary of what is written in the other three biblical Gospels and in the Acts of the Apostles. The verses were probably added during the 10th century. Without these verses the end of the gospel is abrupt, and one cannot tell if there was once another ending that has been lost or if the gospel was meant to end as abruptly as they do without these closing verses. From these verses in the end of the Gospel of Mark in the New Testament of the Bible, people in the eastern United States have gotten the inspiration to catch venomous snakes in the wild, have these snakes as part of their ceremonies and thus risk the health of both animals and congregation members. It also happens that people drink poison during church services, but it is never snake venom. Often it is diluted strychnine, but battery acid also occurs. There are reports of believers dying after drinking strychnine during a church service, but this is not a common occurrence. In addition to snakes and poison, people speak in tongues, practice laying on of hands and holding an open flame against their own skin during these services.

Photos: Friends and family to Cody Coots. Published with permission from Cody. The snake seen in this picture is a Northern cottonmouth, *Agkistrodon piscivorus.* Below you can see Cody holding a flame to his hand.

Surprisingly few people get bitten during church service, a fact that is commonly attributed to some kind of trickery on the part of serpent handlers. Research suggests that a person's chance of being bitten is rather low, especially if they only handle venomous snakes occasionally. Venomous snakes are often calmer than we give them credit for, and they can become accustomed to being held and thus less likely to bite. Other researchers disagree with this explanation, suggesting that the common practice of crowding multiple snakes in a single box leaves the animals stressed which in turn weakens the immune system. This leads to sickness and parasite-outbreaks leaving the snakes listless, unable to muster up the energy to strike. There is some evidence for this line of thought. A group of snakes confiscated from a church in Tennessee were so sick that Michael Ogle, a curator of herpetology at the Knoxville Zoo, was forced to have them put down rather than risk the rest of the zoo's collection. The counterargument is that most of them are freshly caught before every service and released soon afterward. Handling sick snakes would be like handling non-venomous snakes, and that would defeat the purpose. The point is to have enough faith in God to take up something wild, untamed and unpredictable.

However acclimated to humans a snake might be, it is never truly tame. The more often a person handles venomous snakes, the greater the chances for a bite. In my view, sooner or later you have had a venomous snake in your hand one time too many.
Fortunately, bites aren't necessarily fatal. Toxin is metabolically expensive for a snake to produce and the main purpose for the venom besides defense is subduing prey. Snakes rather often give humans dry bites, warning nips that inject none or very little of their precious venom. A full bite from some vipers being used under ceremony, like the Eastern copperhead, is easily survivable. Contrary to popular belief, it is impossible to become immune to snake venom. Instead, successive bites often lead to worsening allergic reactions. [1]

The Pentecostal practitioners are aware of the risks.
There's a constant repetition in recorded interviews: "*Don't take up a serpent if you don't feel the Spirit. Don't handle for show. There's death in that box and you open it at your peril.*"
At the same time, they embrace the risk. A common misunderstanding is that those who practice the faith do not believe they will get bitten by the snakes and that they see death as proof of a lack of faith and a one-way ticket to purgatory. This is not true. Having obeyed God's will by dealing with a venomous snake is said to be an assurance of entering his kingdom.
Life and death rest entirely in God's hands. They believe that whatever happens while handling the snakes is His will. Sometimes He holds back the snake. Sometimes He calls them to Heaven by letting the snake bite. It's not something they expect the rest of the world to understand, and it is something that I can't understand. I believe that the "*don't handle for show*" part was probably a bit overlooked when some pastors became celebrities via reality TV a few years ago.

Pastor Jamie Coots received an almost world-famous snake bite in 2014. Coots was bitten by a Timber rattlesnake on his right hand during service in his church "Full Gospel Tabernacle in Jesus name church" in Middlesboro, Kentucky. After the bite, Coots dropped the snakes but picked them up and continued the ceremony. After the service, he was driven home, despite his health condition. When paramedics arrived, they were not allowed to give him medical treatment as it was not compatible with his faith.

Jamie Coots died in his home. Later in the book you can read more, both about Coots and about Timber rattlesnakes.

Churches do not deny anyone the opportunity to seek medical care if they are bitten. Even though, according to many practitioners, this is against their faith to seek care. It is according to them up to God to decide the fate of the bitten person, not the medical staff. However, they do not criticize anyone who seeks treatment, and they always offer to send for medical help.

Photo: Usman_Khaleel, Pixabay. Timber rattlesnake, *Crotalus horridus*. This species is quite common in Appalachia.

The phenomenon of rattlesnakes during church service has always fascinated me. It seems to me to be both a chaotic and life-threatening environment in these churches during service. Two years prior to this book, I wrote the book "Snake church" which, despite its English title, is a book in Swedish. The book had the same subject as this book, but with a slightly greater focus on the snakes than on religion. After the book was released, I posted a copy to the pastor who contributed and let me interview him. We kept in touch, and he suggested that I should write the book in English. Here we are now,

this is the English version. My idea was not only to translate the Swedish version, but I also wanted to add a little and explore the subject a bit more. The idea was that this time I would focus a little bit more on the church and the belief, while the book will also be about the snake species that occur in these churches. I hope you find the topic as exciting as I do!
Handling venomous snakes in church is not exactly a big city phenomenon, it happens in small congregations in rural areas of Appalachia. The members of the congregations often know each other, and it happens that the pastorate is passed down within the family. The idea is that this book will try to balance between telling the story behind these religious rites and telling the story about the American vipers that are sometimes included in the services.
I discuss the rituals, some important people in the world of religious snake-handling practices and the venomous snakes involved. We must also ask ourselves what consequences the practice has for snakes and humans.

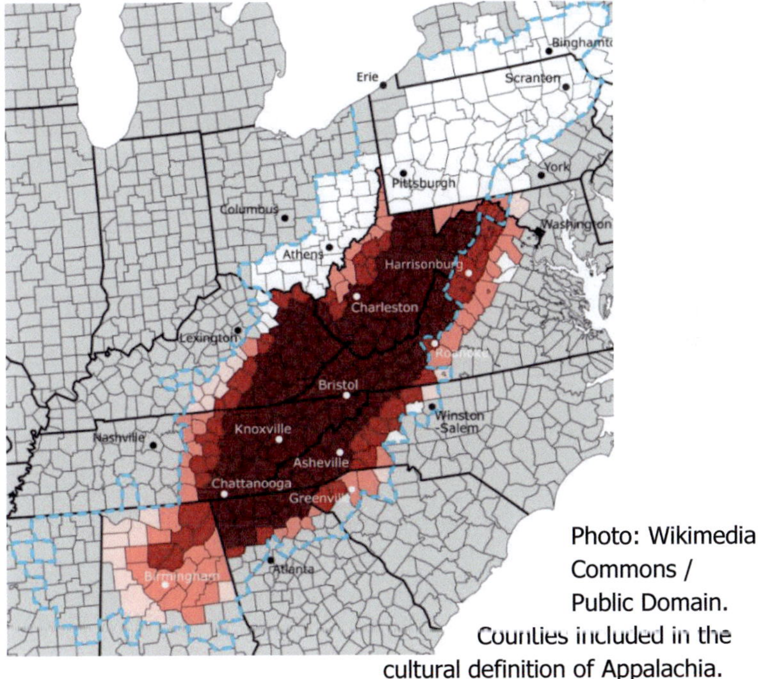

Photo: Wikimedia Commons / Public Domain. Counties included in the cultural definition of Appalachia.

My view.

Before we go further in the book and try to tell the story of this belief and the snakes that are part of the practice, I want to write a couple of short lines about my view on the matter.

I have a small collection of venomous snakes at home, but I choose never to handle these without a snake hook or a similar tool. My belief is that everyone who practices free handling venomous snakes sooner or later has done it one too many times. Venomous snakes must be treated with respect and caution, and accidents can still happen.

Photo: Rickard Ljunggren. Some of my snake hooks. In the background you can see some of my terrariums. All terrariums have a small warning sign with information about the species living in them. If I was to get bitten, I pull the sign off and take it with me to the hospital. It's an extra precaution in case I can't speak properly after a bite.

I am not a believer myself and I find it a bit difficult to accept this literal belief that the book's practitioners have, but my goal with the book is to treat the subject respectfully. On my mother's side, I come from a religious family, but the importance of Christianity in the family has gradually faded. The role of the church has decreased with each generation, we have followed the secularization that has taken place in Swedish society, but we have perhaps been a generation after the great mass who have made the same journey. I am baptized, confirmed and I was married in church, but that was done more because of tradition than faith.

I see nothing strange about someone having a Christian faith, or any other faith for that matter, even though I am not a believer myself. One should not make fun of other people's faith. I think that the phenomenon of religious services where you risk your life is a little crazy and I'm a little curious about who the people are and what drives them to do what they do. It surprises me a bit that somebody can choose to go so far in their practice of faith that they risk their own life and health, or that of others for that matter. For me it is extreme, while it is a bit spectacular with the handling of venomous snakes.
I am also aware that the phenomenon comes from the more rural parts of Appalachia*, in the parts of the United States where the inhabitants are often called hillbillies or mountain people. I'm not going to make fun of the people of the region, I just want to tell the story of the snakes, the people and the church and maybe add my opinion here and there.

*Appalachia is a cultural region in the eastern United States that stretches from southern New York state to northern Alabama and Georgia. While the Appalachian Mountains range stretches from Canada to Alabama, Appalachia usually refers only to the cultural region in the central and southern parts of the range.

Summarized: My point is not to make fun of religious people or people from eastern USA. I think that what they do is dangerous and a bit crazy, but for me there is a difference between (in the eyes of others) doing something crazy and being a lunatic. I am writing the book because it is a subject that I find interesting. Snakes and history have been an interest all my life, the subject of the book is simply an exciting combination.

I hope that I manage the balance and that I do not offend anyone.

The churches that practice the handling of venomous snakes during their services are a vanishingly small part of American Christianity. It is a small part of the American Pentecostal church, most people who belong to the American Pentecostal movement do not come into contact with snakes more often than we do in Swedish churches. Services at *the House of the Lord Jesus* are similar to those practiced by other Pentecostal faiths - except that followers here use snakes, fire and poison in their worship. Handling snakes is a religious rite that occurs in some churches in the rural southeastern United States. These churches are part of *the Sanctification movement, the Pentecostalism movement, Charismatic Christianity* or any other of the few movements with a Protestant orientation that combine strong biblical faith with revival piety and an emphasis on sanctification. According to Christian belief, sanctification is the process by which the Christian is transformed into ever greater holiness and likeness to Christ.

Photo: borkertd, Pixabay. The Timber rattlesnake, *Crotalus horridus*, is the species most commonly used during religious services in Appalachia. It is a beautiful creature with a nasty bite.

Rattlesnakes, which probably are the most famous snakes from the United States, are included in these rites. However, the rattlesnake is not a species, but a genus of snakes that is part of the viper family called *Viperidae.* This family of vipers also include the Eastern copperhead *(Agkistrodon contortrix),* and the Northern cottonmouth *(Agkistrodon piscivorus),* two other American venomous snakes used in religious practice.

In 2022 I wrote a book in Swedish about this subject. In the beginning of that book, I wrote that I had found information stating that cobras and other venomous snake species from Asia have been used in these churches but since I had not had it confirmed I did not include that in the Swedish book. I had read about pastors who started visiting reptile fairs to buy Asian cobras and pit vipers, but it was yet to be confirmed. In the end of the work with that book it was confirmed, sometimes the venomous snakes come from Asia or other parts of the world. It is, to believers, not important that the snake being used in the service comes from Appalachia. It is only important that it is a venomous snake, the species or the range of the species is irrelevant.

The snakes used in these churches are often harvested locally around the church. In some cases, the snakes are bought from someone who caught them, but it is common for the pastor or someone in the congregation to look for snakes in the wild. In some states it is illegal to own snakes that occur in the wild, so instead they sometimes buy captive-bred or wild-caught snakes that are found in the wild in other states or in other parts of the world. Most commonly, however, they are snakes from the region.

Between services, the snakes are kept in terrariums, closed wooden boxes or the like. There are repeated reports that these wild-caught animals do not live that long in captivity. Getting wild-caught animals to function and survive in captivity is sometimes no easy task.

Photo: rogerl01, Pixabay.

Each year, an estimated 7,000-8,000 people are bitten by venomous snakes in the United States [2]. Of these, about five people die each year, but the number of deaths would be much higher if people did not seek care. A certain over-representation of the Appalachian religious practitioners covered in the book can be seen. They encounter venomous snakes more frequently than other residents in the country and some of them like to turn to God instead of the healthcare system after a venomous snake bite.

Most fatal bites in the United States are attributed to rattlesnakes of various species, but copperhead bites are also fatal in rare cases. Copperheads are responsible for the most venomous snake bites in the country, but the bites are rarely fatal [3]. Rattlesnake bites are approximately four times more likely to result in death than copperhead bites. The most bites per year occur in North Carolina, followed by West Virginia, Arkansas, Oklahoma and Virginia in descending order. Men are bitten more often than women, and if I'm right it is the same with our vipers in Sweden, men are slightly overrepresented in the statistics. The US statistics also show bites from Black mamba (*Dendroaspis Polylepis*) Monocle cobra (*Naja kaouthia)* and other species of venomous snake kept in terrariums, but these bites account for a small percentage of the total number of bites.

The idea of including venomous snakes during service is said to have been inspired by the Holiness movement. It is a Christian revival movement built on John Wesley's doctrine of sanctification. John Wesley was born on June 17 in the year 1703 in Epworth, Lincolnshire and died on March 2, 1791, in London. He was a British priest and theologian and is best known as the founder of Methodism. Wesley coined the term "entire sanctification", synonymous with Christian perfection. Wesley said that entire sanctification enables people to fulfill the Great Commandments enunciated by Jesus: Love God with the whole heart, soul, mind, and strength, and to love one's neighbor as oneself (Mark 12:30-31).

He was a strong advocate of temperance and opposed the social exclusion that became one of the consequences of the advance of industrialism. He was also strongly critical of smuggling and the slave trade, but probably had very little to do with snakes.
In 1735 he went to Georgia to work as a missionary among colonists and Native Americans. In 1738 he returned to England, probably without being able to guess that he had laid the foundation for religious people to start dancing with venomous snakes almost 200 years later.

Dancing with and handling venomous snakes in American churches originated in early 20th century Appalachia. Dancing with snakes and handling snakes during service only plays a small role in their practice of religion but it is of course what you are familiar with if you are not a member of the Holiness movement or happen to have a great interest in religion. Members themselves say that too much importance is given to the snake handling by outsiders and that they themselves are just ordinary people practicing religion.

The movement's beliefs and handling of venomous snakes have been documented in several articles and films and it has been the driving force behind several state laws related to the handling of venomous animals. Beginning in the 1940s, several southern states passed laws prohibiting snake handling in religious services.
In 1941, Georgia passed a law that made snake handling a felony punishable by twenty years in prison for injury to another or the death penalty if the injury resulted in death. The Georgia law was written after a six-year-old girl was bitten during a church service near Adel in Cook County. The police arrested both her father and the congregation's pastor, Warren Lipham. The law was repealed in the 1960s. Today, the handling of venomous snakes in Georgia is legal only with a permit.
In Tennessee it is illegal to display any venomous reptile in a way that could endanger others, and Alabama has a similar law. In Kentucky, it is illegal to display any reptile at a religious ceremony. Prosecutions and punishments are quite rare, and a lot happens behind closed doors these days. Even those who have sought attention via media, television and social media escaped the long arm of the law. It seems that this delinquency is not a priority. It could possibly also be the case

that it conflicts with the right to practice one's religion and that this right is stronger than animal rights.

Unfortunately, USA does not have as strong animal protection laws as we have here in Sweden, especially not if we are looking at snakes. In Sweden it would be impossible to practice snake handling during ceremony. All snake species are protected in Sweden, in some states in the USA they even make family activities out of catching and killing rattlesnakes. These events involving the killing of rattlesnakes are strongly questioned by animal rights organizations, but so far, they have not succeeded in stopping what they consider being animal cruelty.
These cruel events where lots of rattlesnakes are being killed as entertainment have nothing to do with the church world.

*

"The Copperhead cuddlers" case, an example of law enforcement action and the following trial.

On November 1, 1947, Durham police (Durham is a city in North Carolina) raided *the Zion Tabernacle Church* on Peabody Street and confiscated a copperhead snake that was being handled in a church service. The night of the raid, the church was filled with congregations' members that were dancing and singing as police came inside and used a pronged tool to put the snake in a bucket, The Durham Morning Herald reported Nov. 2, 1947.
While no arrests were made that night, during the next several days, individuals, including Reverend Colonel Hartman Bunn, were reminded of a city ordinance banning the handling of venomous reptiles. No arrests were made pending examination of the snake to determine if it was venomous or not. An expert at what's now the *N.C. Museum of Natural Sciences* verified that the snake was indeed a copperhead.

Bunn and other church members continued to handle snakes at ceremonies on subsequent nights, vowing to take their fight to the Supreme Court. After he was arrested, Bunn pleaded not guilty in court before a standing-room-only crowd and asked for a continuance.

On November 19, Bunn and his snake carer, Benjamin Ralph Massey, were found guilty and fined. "*The lower court ruled that snake handling endangered public safety and fined the two $50 each and court costs. The snakes were tested, the court noted, two healthy rats, which were placed with them in a cage, were struck and died almost immediately. Bunn claimed that snake handling had been practiced for 40 years in North Carolina without harming anyone.*
Chief Justice (Walter P.) Stacy wrote that the case rested on "a very simple question: which is superior, the public safety or the defendants' religious practice?" The court found, despite Bunn's apt argument, that the case was simply a question of snakes or people. The people won."

Both appealed to Superior Court. Though the church leader pushed back on the grounds of religious freedom, the state Supreme Court upheld the original decision. Bunn made an impassioned speech on the steps outside court, in which he compared the police who confiscated his snakes to Hitler's SS Elite Guard. Quite harsh words today, but two years after the end of the Second World War, it can be seen as a sulfurous accusation.

Life Magazine became interested in Durham's snake-handling congregation, and after learning of a snake-handling convention to be held in *Zion Tabernacle* in the fall of 1948, it obtained permission to take pictures. The magazine later published some sensational photographs: copperheads and rattlesnakes freely wrapped around a visiting preacher's head or held by members of the congregation while people hummed, chattered, spoke in tongues, danced, and generally hypnotized themselves into a religious frenzy in which faith cast out not only fear but reason. The congregation's interracial makeup, the character of Bunn—mild-mannered, soft-spoken, unexcitable-and the unfamiliar behavior of the participants, workers in the cotton mills, the tobacco factories, and Wright Machinery Company, utterly confounded the authorities. The snakes were again confiscated, and arrests were made. Bunn appealed the charges and challenged Durham's ordinance on the grounds of its interference with religious liberty, but the State Supreme Court upheld its legality.

Photo: State Archives of North Carolina Raleigh, NC.

This photo was taken during the two-year anniversary (1949) of the raid, and it shows churchgoers with snakes in their hands and around their necks. No one was injured during the two services that were observed by the police.

On the far right you can see Reverend Colonel Hartman Bunn with snakes in his hands. Later Bunn left the city and when he returned to Durham and *Zion Tabernacle* in 1955 and broadcasted once a week over WTIK, snakes were no longer a part of his service. He died 1991, at the age of 81. The building where Bunn had his services was demolished in 2008.

In the picture you see a black woman with a Timber rattlesnake in her hands. This is one of only two pictures I have seen that does not show only white people in the churches that fit within the scope of this book. I don't know why this is the case, but I have some guesses. Many images are from the era of segregation, and perhaps it was mostly whites who were attracted by/gained access to congregations where the snakes were used during worship. It is also the case that the few film clips from church services that you find on YouTube and similar

show only white congregation members. It is a question that I will try to answer later in the book, I think that it is an interesting question for the interviews.
It is easy to dismiss the members of these churches as crazy hillbillies and perhaps there is something to that. Can one be sane and still participate in this spectacle? Is it reasonable to risk one's life in the name of God? I am not a believer, but for me others are welcome to have their faith, regardless of which god they worship. However, I have always found deep faith to be a bit alienating and sometimes almost a little scary. Whether they speak in tongues, dance with venomous snakes, persistently try to push their beliefs on to others or mass as if in trance. Overly intense people gaping and flailing their arms feels a bit awkward and scary to me, especially if it's done in the name of religion. It is of course even more frightening when believers have the conviction that one should harm others in the name of their god, but that is outside the scope of this book and nothing that these practitioners can be accused of. Those who dance with or handle snakes during service may hurt themselves and possibly other church members, but they do not wish you or me any harm. They don't really want to hurt the snakes either, but their practice sometimes harms the snakes being captured to be used as props.
Unfortunately, I have seen a video that shows a real misunderstanding of snakes and their welfare. I saw a clip where a grown man stands on a rattlesnake and remains standing while he is massing to the congregation. Of course, no snake is made to support the weight of a full-grown man. The reporter who witnessed the incident asked the pastor afterwards if standing on it did not hurt the snake. In response, he received a horribly wrong exposition on the anatomy of snakes, with the conclusion that snakes could not be harmed by standing on or stepping on them.
My thoughts: Generally speaking, snakes naturally do best when left alone. For the snakes, a church service means full chaos and enormous stress. A snake is best off having as little as possible to do with us humans. Contact means increasing stress for the snake, a greater stress if it is a wild-caught animal even though snakes born in captivity are not domesticated the way a dog is. In my opinion, it is animal cruelty, regardless of whether you choose to step on the snake or not.

The believers handle snakes because it's safe.
It says so in the King James Version of the Bible, Gospel of Mark, chapter 16, verses 17 and 18.

"And these signs shall follow them that believe; in my name shall they cast out devils; they shall speak with new tongues; they shall take up serpents; and if they drink any deadly thing, it shall not hurt them; they shall lay hands on the sick, and they shall recover."

They call themselves *signs-following* churches, those whose belief is proved through their tongues speaking, poison drinking, snake handling, and faith healing.
Deaths from venomous snake bites have occurred since the congregations began handling snakes and are still occurring.
If you compare the number of practitioners to the number of inhabitants in the United States, there is an overrepresentation of snake-bitten people in the Pentecostal churches covered in this book. Perhaps it is not so surprising, most people living in the US probably encounter snakes quite rarely and handle snakes with their hands even more rarely. More than a few pastors and some parishioners in snake-handling churches get bitten. Many of them survive with minor injuries, but a bite can be fatal. The number of documented deaths due to venomous snake bites in the church world varies, from 90 to 120 from the first documented death in 1922 until today. The practice does not seem to present a danger to observers. At least there is no documented case of a non-handling member being bitten by a snake handled by another believer.

A pastor named J.E. McCora was the first to die from snake handling, at least this is the earliest case I have found. In 1922, during a revival meeting in the Cherokee County area of North Carolina, McCora was bitten by a rattlesnake while performing his practice.

In 2015, John Brock died after being bitten by a rattlesnake during a service at *Mossie Simpson Pentecostal Church* in Jenson, Kentucky. Brock, a preacher from Stoney Fork, was bitten in the left arm during service. He refused treatment and died in his brother's home. This appears to be the latest documented death and hopefully the last.

Just because these bites are rarely fatal if you have the sense to seek care, it doesn't mean they aren't extremely painful bites to experience. Depending on the species, snake venom attacks the central nervous system, but also causes severe pain, convulsions, tissue damage, paralysis, nerve damage and organ failure. It's not like in the movies, James Bond gets a shot of serum, get back up on his feet and catches the bad guy. It can take time to recover from a venomous snake bite and you can suffer from permanent impairments or injuries. Operations and tissue damage can lead to ugly scars. It is not unusual for pastors and parishioners who have been bitten to choose to forego treatment. If you have the right faith, you will survive the venom... This is probably part of the idea of dealing with venomous snakes in church, by not getting bitten or by surviving a bite, you show yourself and your congregation that you have the right and pure faith. One has attained entire sanctification and now has proof of it.

In recent years, however, there has been a change. Nowadays, many younger practitioners interpret the text of the Bible to mean that one can seek medical care if necessary.
Ralph Hood, a professor at the University of Tennessee who specializes in the psychology of religion, said in an interview that refusing to call 911 for help is now considered "old school." Younger pastors claim that no verse in the Bible forbids anyone from seeking help after a venomous bite. You simply seek help from both Jesus and the nearest hospital. Older snake handling pastors like Jimmy Morrow (snake-handling pastor of the Edwina Church of God in Jesus Christ's Name, in Newport, Tennessee) believe in sticking to their traditional way of practicing their faith.

"I've been bit twice by a Copperhead, and I didn't go (to a doctor). *I just stayed home, and the Lord healed me. I know a lot of good brothers and sisters who say that when they die, they want to die while practicing the signs of the Gospel."*

The snake care and the criticism surrounding this.

A large proportion of the snakes used by the parishes are taken from the wild. Snakes are also bought, both those caught by others and those bred in captivity. However, it is most common that the snakes come from the parish's nearby nature. I saw an old interview where Andrew Hamblin told TV viewers that he only collected males from the wild. Since Andrew (more about him later in the book) didn't want to be interviewed, I asked Cody Coots the questions instead. Whether it is common for parishes to collect only males from the wild and if so, why they do so. His answer confirmed what Andrew had said.

"We keep the females in the wild that way they can have their baby's. Here's my dad's rules he gave me about hunting:

1. *No females*
2. *No rattlers under 3 ft*
3. *No copperheads under 2ft*
4. *Never hunt the same spot more than once a month.*

Some people bring females and small snakes out but that kills the population.
And this is a no-no in Kentucky but when I'd buy canebrakes and southern coppers in bulk, I'd turn the females lose. Over the years I've found cane/timber crosses. You could tell by the stripes in the face."

Snake handlers usually keep their animals in terrariums or boxes in sheds, feeding them live mice about once a week. Critics of this religious practice argue that the snakes are often dehydrated and malnourished, making them less likely to bite and thus less dangerous to handle. Zoologists have examined fecal material from some congregation's snakes and concluded that sometimes the snakes are malnourished and dehydrated. Sometimes no fecal material can be found in the snake containers, which indicates that the snakes are not being fed. The animals are often treated poorly, according to Kristen Wiley of the Kentucky Reptile Zoo, which sometimes ends up caring for—or euthanizing—weak and dehydrated snakes confiscated from Pentecostal handlers.

The famous pastor Jamie Coots, who also became something of a reality star, denied these claims but has himself said in a radio interview that his snakes lived 3-4 months on average, and that many of his snakes did not want to eat [4].

Danny Hobart was asked in an interview if the snakes in the congregation were ever tame, so that the risk of bites was reduced [5]. His answer unfortunately confirmed that snakes are often not long-lived in the Pentecostal world.
"-Danny, does it ever happen that the serpents get used to you? And after a while they don't mind people holding them?"
"-No, you can't tame a serpent. You might keep one maybe, if they could live a year, and it would still bite you. You'll not tame them."
"-What do you feed them?"
"-Feed them? Most time when you catch them, and they in captivity like that, they won't eat."
"-They won't?"
"-They eat dirt. Yeah."
"-They do? I didn't know that."
"-Now, sometimes we have force fed them before."
"-What do you feed them?"
"-What didn't we feed them? You can get mice and feed them, or raw meat. They'll eat raw meat. Most time when you get them in captivity like that, they won't eat though."

I have seen videos of snakes being fed mice; in some congregations they manage to get at least some snakes to eat. It is impossible to say what percentage of snakes or parishes this applies to, what can be said is that in some clips you see it is almost lifeless snakes that are handled in the church.
The two pastors who agreed to be interviewed for this book claim that their snakes are well cared for. One of them is known for taking good care of his snakes and I have seen video footage of the other one's snake room. It was not the best reptile keeping that I have seen, but I have seen worse in the reptile hobby more than one time.

My opinion is that even with good husbandry in decent terrariums, the environment and handling in the church is far from optimal for snakes. Snakes are animals that tend to avoid humans. Being exposed to rough handling, with a lot of people around, must add to the snake's stress levels. Add to that the loud volume of music or preaching and smells from everything from people to venoms and open flames. The fact that some practitioners take off their shoes, and sometimes their socks, and stand on a snake is unimaginable to me. I don't know how common this phenomenon is, I have come across three pastors who have been filmed doing it.
In the reptile hobby in Sweden, there are often complaints about beginners who pick out their corn snakes / ball pythons in time and time. The snakes do not do well if they often spend long periods outside their terrarium, where you have hopefully tried to create an environment that is good for the species. As pets, they are mostly animals you don't socialize with, they are best left alone in their terrariums. The same should apply to venomous snakes in the US and probably more so if they are wild caught. It is also the case that the environment of a reasonably crowded church offers snakes more stress than a reptile owner's home does.

I have chosen to include in this part of the book a few critical opinions, from people who have been on the spot and seen the handling or who have worked with taking care of seized snakes.
Kristen Wiley was easy to find. In the autumn of 2024, there is a heated debate on social media about a free handler who was bitten by an Inland taipan *(Oxyuranus microlepidotus)* while seeking fame on social media. Kristen became involved in the following debate, which reached much of the online reptile community.

I contacted her and on the following pages you can read my mail and her response.

Hello!

My name is Rickard Ljunggren, and I am vice-president of the Swedish Herpetological Society. In my spare time I write some books on the subject of herpetology. I'm currently writing a book about the Pentecostals who use snakes during worship and your name has come up on a couple of occasions, as someone who has had to deal with seized snakes. Would you like to tell us a little about your views on the subject? Do you think that snakes are well cared for by churches and pastors or is it neglect? Maybe both, depending on the parish? What condition are the snakes usually in? Is it sometimes the case that snakes are not confiscated because animal husbandry is good? What do you think about the handling of snakes in the church, from the perspective of the snakes?

Best regards Rickard

Kristen replied quickly and wrote to me that she would gladly answer follow-up questions.

"Hello Rickard,
Thanks for sending an email. In my experience, many of the snakes kept by the religious snake handlers in the American southeast are in very poor condition.
People who handle in church are not ignorant of snake behavior. They understand that the snakes are reluctant to bite and use their natural inclinations to get away with free handling them. In general, the snakes are either handled in a way that simply keeps them off balance and is relatively gentle; or they are handled in a way that is highly disorienting and creates an extreme stress response in the sakes. This second method can involve techniques like flailing the snakes like they are whips, shaking them vigorously, or dangling them from a single point while swinging them around. Both techniques result in fewer bites. In addition, many snake handlers do not take adequate care of the snakes. Wild-captured snakes are frequently not given any time to acclimate to captivity before being used in services. Little to no effort is made to get them feeding and established before they are subjected to the higher stress of a church service. Sometimes snakes are held with no water for extended periods of time. All of these factors result

in snakes that are much less likely to bite and are much less likely to inject a significant amount of venom if they do bite. When we received confiscated copperheads several years ago, over 100 individuals were able to give only a few drops of venom in total.
Snakes are so gentle and shy by nature that people are able to take advantage of them in this way. There is absolutely nothing supernatural about Pentecostal's ability to handle snakes with few bites and few deaths. In the same way, there is nothing magical or special about the many individuals online now who free handle for views. I'm happy to answer any further questions.

Best, Kristen"

The second person I contacted was Steve Ludwin, who had filmed a short documentary about his visit to one of the churches where venomous snakes are part of the service. It was Chris Wolford's congregation he visited and his sermon that Steve took part in. You can read more about Chris Wolford later in the book. I asked Steve to write a short introduction for you readers, as he is quite a multifaceted person. He describes himself as a published musician and snake venom researcher with keen interest in anti-aging and longevity. I agree with his presentation, and I would also like to add that he is a reptile enthusiast and that he is quite active on YouTube. He has a collection of snakes himself, both venomous and nonvenomous, and he is well familiar with snake care. I chose Steve as a critical voice because he has seen snake handling during service and because after the sermon, he took a discussion about what he considered to be animal abuse. You can find this documentary on YouTube. Steve Ludwin is also known for milking venom from his snakes and injecting himself with it. From vipers to cobras, the snakes that have donated venom for Steve's experiments have pushed his body to the limit. He has even been hospitalized after mixing and overdosing on the toxic substances.

This is his story about the incident:

"I had one near-death experience in 2008, when I overdosed. I injected myself with a cocktail made with venom from my Crotalus oreganus, Bothriechis nigroadspersus and Trimeresurus albolabris. Three different continents and that was a real problem. I did it as part of a health experiment, but it went horribly wrong. I did not mean to inject as much as I did, and when I did it was all over. I put the needle into my left wrist, and as soon as the venom went in, I knew it was game over. My hand swelled up like a baseball glove and my arm filled with fluid all the way up to my shoulder. My view afterwards is that it was a very foolish thing to do. When I told the doctors that I had deliberately injected three different deadly snake venoms, they couldn't believe it. I spent three days in intensive care; they said there was a strong possibility they would have to cut off my arm. In the end, they didn't, and I discharged myself. A week later, they insisted that I must go back and see them. They took photos of my arm and told me they had never seen such a recovery. I have no doubt that this was because of all my previous venom injections. I believe that if I were to be bitten by a venomous snake, my chances of survival would be very strong."

During October 2024, Steve and I chatted a lot via WhatsApp, he answered my questions and emailed me pictures. Now it was time for a phone call! I was really curious about his visit to Chris Wolford's church (look it up on YouTube!) and wanted to hear him talk about his experience. It was quite a long phone call where we talked about everything! About snakes of course, but also about his research on snake venom, Satanists burning stave churches in Norway in the 90's, religion in general, free handlers as a social media phenomenon, music, Denmark and a little bit about everything. Steve is a multi-faceted and pleasant person with opinions on many things, he has a standing invitation to my home. I wrote down Steve's story about the trip to West Virginia he took a few years ago:

"I am an atheist myself, born Catholic but left the church world as soon as I realized that dinosaurs were real. My opinion is that the bible is just a book and that a lot of animal abuse against snakes is because of this book. My opinion on free handling is that it is okay, but I think

these churches should have better terrariums for their snakes. I myself have been bitten three times by venomous snakes. Twice by eye lash vipers and once by a copperhead. I sometimes free handle myself, but I don't do it to chase likes on the web. I am from the USA but had not been there for many years when I travelled to visit the church that was expecting me, I live in the UK. I had heard about the phenomenon of the rattlesnakes in the churches for many years, but this was my first (and only) visit. It was a long journey and apart from the flight, we drove for 5-6 hours to get to the church which was way out in the countryside in West Virginia. The whole experience was a bit crazy! My first impression when I entered the church was that there was a lot of tension in the air. Maybe it was because I was there? Anyway, everyone was very nice and friendly to me, they offered me food that was prepared in the basement. The congregation was a mixed group of people, perhaps with a predominance of older women. Some of them held flames to their bodies, some drank strychnine. There were also some former addicts in the congregation, who had found God. There was very loud music playing, my experience was that it was a bit chaotic in the church. The snakes were only there for a short part of the service. The whole service probably took 2 or 3 hours, but the snakes were only there for about 15 minutes.

I was not at all comfortable being there and I was upset to see snakes being mistreated in the name of God. My immediate reaction when Chris Wolford stood on a snake was that I wanted to grab the snakes and run out of the church with them! I was angry and walked out of the church! I think their handling of snakes was okay in general, when they maybe danced a little. However, it was not okay when they swung the snakes, which were then completely straight in their body, like a rope being swung. If nothing else, that probably disorientated the snake quite a bit! After Chris Wolford swung a Timber rattlesnake, he put it on the floor and stood on it! Incredibly cruel! Snake abuse in the name of God! I got the impression that the snakes were cooled down, that the handling is a circus trick based on the fact that cooled down snakes are a bit sluggish and that they then bite less often. The snakes were lethargic, but I could not see that they were dehydrated or malnourished. I also got the impression that they knew very little about snakes, otherwise they would not have stood on a snake and afterwards tried to explain to me that the snake is not harmed by this.

My opinion is that venomous snakes should be banned from churches, just as rattlesnake roundups should be banned."*

This was Steve's story. He still gets upset when he talks about what he saw and the lack of understanding he encountered when he tried to talk about it outside the church afterwards.

*Rattlesnake roundups are annual events common in the rural Midwest and Southern United States, where the primary attractions are captured wild rattlesnakes which are sold, displayed, killed for food or animal products (such as snakeskin) or released back into the wild. My opinion is that venomous snakes should be banned from churches, just as rattlesnake roundups should be banned. It's a family entertainment that is largely about seeing rattlesnakes killed and perhaps having the pleasure of doing it yourself. You can take selfies with rattlesnakes with their mouths sewn shut, there are beauty contests (who wouldn't want to see a pretty girl kill snakes??) and prize-giving ceremonies. A harvest of several hundred to several thousand kilograms of snakes is typical for many roundups. In Texas, up to 125 000 snakes could have been removed annually from the wild during the 1990s. The largest rattlesnake roundup is held in Sweetwater, Texas. Held annually in mid-March since 1958, it now turns over several million dollars each year. These events have, as I stated before, nothing to do with the Pentecostal churches. Some roundups have transitioned to no kill rattlesnake festivals which still draw crowds focusing on education and the economic benefits of rattlesnakes. These festivals are something worth celebrating.

2. Appalachia

Photo: dafacct, Pixabay. Grist mill, Babcock State Park, West Virginia.

The movie *Deliverance* from 1972 takes place and is filmed in the Appalachians in Georgia. The film portrays the people from the region as inbred, backward and dangerous and portrays the region as very poor and far behind in development. The film has had an enormous impact, and it has probably largely shaped the world's view of hillbillies. Although the film is over 50 years old, many people are familiar with the infamous "squeal like a pig" scene.

The 2012 documentary *The Deliverance of Rabun County* explores how the film affected the people of the region and how they felt about how they were portrayed. Many of those interviewed for the documentary felt resentful of the way they were portrayed.

The *Wrong Turn* horror film series consists of seven films released between 2003 and 2021. Each film in the series takes place in a different location in rural West Virginia and follows the story of a group (a new group of people in each film) of travelers who get lost in the woods. Stereotypes of the people of Appalachia are mostly depicted in the films as inbred and cannibalistic monsters that hunt and kill the group of travelers. These are two of many examples of how movies portray people from Appalachia.

Photo: Peter Adair, 1967. Wikimedia Commons / Public Domain. Still from his documentary *Holy Ghost People*.

Holy Ghost People is a 1967 documentary directed and narrated by Peter Adair. It is about the service of a Pentecostal community in Scrabble Creek, West Virginia, United States. The church service in the documentary includes faith healing, snake handling, speaking in tongues and singing. You can find it on YouTube!
(At least in the summer of 2024 when I am writing this chapter.)

The documentary begins by showing images of the church and its night services. After the opening credits, a narrator introduces the Pentecostal community in Scrabble Creek. The narrator presents various activities the church partakes in, such as snake handling, speaking in tongues, and four-to-six-hour long meetings at the church multiple times a week. The narrators explain that while people are often bitten while handling the snakes, mainly copperheads, they refuse medical help.
The documentary then features several interviews with members of the church. These interviews reveal stories of how many of the church's members found salvation through the Holy Ghost and how

the Holy Ghost saves them in their daily lives. Some members reveal stories of how they are able to speak in tongues, others reveal how they communicate with God. The final interview is of an old woman who shakes and sometimes convulses on camera while going in and out of speaking in tongues.
The film then cuts to the beginning of a church service. When everyone is seated, people start clapping and singing together. Then there is a cut to the inviting of those who have not found the Holy Ghost to find out. He also tells the congregation to ignore the cameraman and to act as though it was just another normal night. Eventually, the church service moves into a time of prayer. The people stand and announce their prayer concerns to the congregation. The pastor tells people that God will answer their prayers if they only believe. They then bring a woman up front, who is rapidly losing her eyesight, for the congregation to pray for her. The camera pans to the rest of the church and shows that everyone else has formed small groups and they are all praying at the same time. The different styles of prayer include standing still, lying on the floor, and convulsing seemingly uncontrollably.
The end of the film contains a lot of fast cuts to show everything that happened in the service. A new man preaches, followed by two people who lead the congregation in worship. After clapping and singing, snakes are brought out to the snake handlers. As the music and clapping continue, people begin to get up and dance. Various people throughout the church handle several snakes, and a man who dances violently quickly collapses to the ground and lies there. The music stops so that people can provide testimonies, and the church can take an offering. The pastor handles a snake as he tries to get people to give money to the church. The snake bites the pastor on the hand and the film ends with a shot of the pastor's swollen hand.

The picture quality is lousy, I can't really identify the species of the snake that bit him. Even though the picture quality is bad, it is worth a watch. The documentary gives a good insight into the services and gives me the impression that it is true that snake handling is only a small part of the service. The documentary was a partial inspiration for the 2013 film *Holy Ghost People*, and some of its footage was used.

Holy Ghost People (2013) was meant to be an even-handed portrayal of Pentecostal Christians, and the film was written to explore both the positive and negative aspects of faith. The film was shot in Tennessee at Camp Nakanawa. In an interview, the director said he wanted to *"get back to [his] roots and tell a gritty American tale."* Real snakes were used, and the snake handling actor prepared for his role by watching videos of snake-handlers, whom he described as *"charming and funny and charismatic, they are performers."*

In recent years, the world has gotten an updated view about how hillbillies are. A family known as "the most inbred family in America" has become famous after a documentary on YouTube.
Mark Laita, who made the documentary, didn't get a friendly hello when he first showed up to meet the family in 2004, with a protective neighbor threatening him with a gun.

"They are kind of protected by the neighbors and the relatives (who) *don't like these people coming to ridicule them."*

But Laita became friends with the Whittakers and stayed close to them for 20 years.
In 2020, Laita made his documentary about their life in 2020, which since then has been watched more than 30 million times. He returned in 2022 for another follow-up after setting up a GoFundMe page, which has raised thousands of dollars for the family.
On the *Koncrete KLIPS* podcast, he said his first time seeing the family felt like something out of the movie Deliverance.

"We came around to this road, which turns into a country road, which turns into a dirt road, and we come to this trailer and then a little shack on the other side of the road. And there's these people walking around, and their eyes are going in different directions, and they are barking at us. And then one guy, you would look him in the eye or say anything and he would just scream and go running away, and his pants would fall around his ankles, and he would go running off and go and kick a garbage can."

The Whittakers, who live in the small village of Odd in West Virginia, have a complex history. Some of the family can only communicate through grunts and barks and often flee when outsiders try to speak to them. They can struggle to remember their parents and other relatives and didn't realize until recent years that their issues were due to inbreeding. The cycle of inbreeding began with identical twin brothers, Henry and John Whittaker. Their children married each other and had over a dozen children. John had nine kids with his wife, and first cousin, Ada Riggs, including Gracie Irene Whittaker in 1920. Henry had seven children with his wife Sally, including John Emory Whittaker, who was born in 1913. John Emory and Gracie married in 1935 and had 15 children. The couple were double cousins, sharing both sets of grandparents, and their offspring suffered severe physical and mental defects, suspected to be because of their inbreeding.
Betty, who was born in 1952, is the current head of the family.
She, along with her siblings Ray and Lorene, were interviewed in the documentary. Ray and Lorene, who have a son named Timmy, cannot talk properly and make noises instead of using words, but they are still able to understand one another.
It's a sad story, the picture I get from watching the videos on YouTube is that the family lives in misery. Unfortunately, I think the story of the Whittaker family distorts the image of what the people in the region are like, even if that is not the intention of the documentary filmmaker.

If you watch reality TV, there are a lot of shows that take place in the Appalachian region. Some examples are *Moonshiners, Snake Salvation, Hillbilly Handfishin'* and *Swamp people*. What all series have in common is that it is not exactly ordinary people who get a place in the series, it is people who more or less live a life outside society with partially their own rules. I understand that it doesn't make good television to film completely ordinary people living completely uninteresting lives, but this one-sided portrayal in film and television perpetuates old prejudices about what people in Appalachia are like. Of course, it's just movies and TV and it's just entertainment and shouldn't be seen as something that describes the truth, but somewhere we got our idea of hillbillies from. Few of us have been there and met these people, but many of us have a definite opinion about these people. Many have a negative view of Appalachia and its residents because of the "Appalachian stereotype". The generalizations that people living outside the region have about the place and its inhabitants are that Appalachians are lawless, violent, poor, rundown, and uneducated.

*

Hillbilly is a (often derogatory) term used for people who live in rural, mountainous areas of the United States, including Appalachia where the phenomenon of venomous snakes in church is most common. People from the region are largely perceived to be poor, white, rural and unsophisticated. The stereotypical portrayal of the people from there is *"children in light gray-brown clothes with dirty faces, weather-beaten women sitting and smoking on the steps of the caravan and men with alcohol problems and a propensity for violence".* The people from the region are expected to be dirty, something that may have originated from the large number of coal miners in the past. Teenage pregnancy, low education and inbreeding are part of the prejudice against those who choose to call them hillbillies. Those with Appalachian accents or Appalachian dialect are perceived as low-educated and poor. As a result of these negative stereotypes, thousands of people from Appalachia face judgmental scrutiny on a daily basis. The people of Appalachia are seen by the outside world as hillbillies or rednecks and these terms often appear in comical usage. Discrimination against people from the region has been so significant that in some place's laws have been passed making it illegal to discriminate against people with an Appalachian identity.

The term "hillbilly" spread in the years after the American Civil War. At this time, the country was developing both technologically and socially, but Appalachia began to fall behind. Before the war, Appalachia was not distinctly different from other rural areas of the country. After the war, the people of Appalachia were perceived as backward, quick to violence, and inbred in their isolation. Due to news stories of hillbilly feuds, the hillbilly stereotype developed in the late 19th century. The first known examples of the word "hillbilly" in print are from the years around the turn of the century. A *New York Journal* article from the time included the definition *"a Hill-Billie is a free and unrestrained white citizen of Alabama, who lives in the mountains, has no means to speak of, dresses as he can, talks as he pleases, drinks whiskey when he can get it and fires his revolver when he feels like it."*

The classic hillbilly stereotype reached its current character during the Great Depression. The period of emigration from the Appalachians, roughly from the 1930s to the 1950s, saw many mountain dwellers moving north to the industrial cities of the Midwest. This movement to communities in the north, which became known as the "Hillbilly Highway", brought these previously isolated communities into mainstream American culture. In response, poor mountain people became central characters in newspapers and eventually in films. The stereotype is multifaceted as it contains both positive and negative features. Hillbillies are often seen as independent individuals who resist the modernization of society, but at the same time they are also defined as backward and violent. Scholars argue that this duality reflects the fractured ethnic identities of white America.

Photo: Wikimedia Commons / Public Domain. Dolly Parton 1965, image cropped.

The term hillbilly is used with pride by a number of people from the region including famous people such as singer Dolly Parton. Positive self-identification with the term includes stated "hillbilly values" such as love and respect for nature, strong work ethic, generosity to neighbors and those in need, family ties, self-reliance, resilience, and a simple lifestyle.

The Hillbillies were not the first to settle in the region. The indigenous people began to settle in the Appalachians about 16,000 years ago. The Cherokees were the main tribe in the southern Appalachian and Blue Ridge regions, but there were also the Iroquois, Powhatan, and Shawnee. The arrival of enslaved Africans in the area dates back to the 16th century. The indigenous and the African people both had a huge influence on the culture of Appalachia. In the 18th century, settlers came from Europe. Up to 90% of these were Protestants who came from the Scottish Lowlands and the province of Ulster in Ireland. Immigration from these regions continued throughout the 19th century and it was largely poor people who were used to being self-sufficient who came. They had an innate distrust of power and government after decades of fighting the English and Catholics. This is where the cultural stereotypes of family loyalty, rebellion against authority and passion for self-defense arose, which later evolved into the image of hillbillies as wild and withdrawn mountain men.

German immigrants were another group that had a major influence on the culture of Appalachia. These were often called Dutch because they came from "Deutschland". They settled primarily in Pennsylvania and Virginia, bringing with them foods such as applesauce and sauerkraut. Their cultural identity was so strong that they did not assimilate very well. Instead, they often had their own German schools and churches. The Germans were treated better than the Scots, Irish and Italian immigrants, mainly because they looked more like the British colonists. The immigrants from Scotland and Ireland tended to keep to themselves and were generally too poor to own slaves. The Germans had slaves from Africa and their settlement had a much more direct impact on the displacement of the Native American population. The European immigrants claimed lands from the coast west into the Appalachians, many who came moved deep into rural Appalachia. The Europeans' growing need for land led to countless battles against the tribes that already lived there. Eventually, treaties were made with the tribes, unfortunately these treaties ended with almost all of the Cherokee and other native groups from the region being forcibly moved westward.

A small clique of Scandinavians found their way to the region, especially people from Finland and Sweden. These brought woodworking skills from Northern Europe, giving rise to the log cabin.

Appalachia is often considered a rural area populated by whites, but African Americans have inhabited the area for hundreds of years. In fact, in 1860 an estimated 10% of Appalachia's population was black.

As white settlers moved into the region, so did Africans, both free and enslaved. In America's early pioneer era, white, black, and Native Americans all lived close together in the Appalachians. This gave rise in the early 19th century to a people group known as the Melungeons, who had roots in Africa, Europe and North America. Appalachia consisted of a complex mix of ethnic groups, the common trait that united them all was that they were accustomed to hard work and self-sufficiency. They had what it took to survive in the wilderness of Appalachia.

Those of the white elite, and some Cherokees, kept Africans as slaves in southern Appalachia. But the mountain landscape did not lend itself as naturally to large plantations as the areas further south, and the fact is that the majority of the people in the region were not slaveholders. There was an active Underground Railroad that ran through Appalachia, from Chattanooga north to Pennsylvania. The Underground Railroad was an organized network of contacts maintained by opponents of slavery (so-called abolitionists) in order to help slaves escape from the southern states to the northern states or to Canada. The abolitionists offered the refugees help, information and hiding places. Escape routes were called tracks, hiding places stations, refugee passengers and those who helped refugees were called conductors. The slaves knew that slavery was forbidden in the north, and many dreamed of freedom and chose to escape. If the escaping slaves were captured, they were punished by whipping, mutilation, or death. In the southern states, slave patrols were organized with hunting dogs that hunted these people who were on the run.

After the Civil war (April 12, 1861 – May 26, 1865), many freed slaves moved to Appalachia. The African influence on Appalachia persists even today. The banjo—a stringed instrument central to bluegrass and other forms of Appalachian music—originated in Africa. They also introduced foods such as sorghum, sweet potatoes, black eye beans, watermelon, and peanuts into Appalachian cuisine.

The brutality of the Civil war only served to intensify the resentment that many in the countryside had against authorities and outsiders. Many came with hopes of becoming rich but lived a life of poverty.

Miners were paid per ton of coal produced instead of hourly wages. Because of this, many had a very low standard of living and the region's economy did not flourish. The struggle against the low wages resulted in class tensions that ended in violent conflicts between miners and coal mine owners.

Photo: Mathew Brady, picture taken 1862. Wikimedia Commons / Public Domain. Runaway slaves, photographed at Fort Lafayette, Valley Forge, Pennsylvania. In other words, they managed to escape.

*

The increased population growth was a result of the expansion of coal mining which attracted many to the region. Gradually, dissension grew between the wealthy elite living in the lowlands and along the coast, and the more rural people in the mountains and wilderness of Appalachia.

Widespread poverty in Appalachia gained attention in 1940 when James Still's novel *River of Earth*, documenting Appalachia during the Great Depression, was published. By then, the people of Appalachia had already lived in poverty for many decades.

Photo: Russel Lee. Wikimedia Commons / Public Domain. Workers coming out of the mine at the end of a day's work. P V & K Coal Company, Clover Gap Mine, Kentucky, 1946.

In retrospect, it seems that the original settlers' core values of freedom, self-reliance and a unique individual identity eventually brought them into conflict with the progress of modern life. Isolation and a fear of losing touch with their traditional values led to devastating poverty. Into the 60s and 70s, many people in Appalachia still lived without basic necessities like electricity or running water indoors. Hunger and lack of basic hygiene were not uncommon.

In 1965, after President Lyndon B. Johnson declared a "war on poverty" and the Appalachian Regional Commission was launched. The commission focuses on helping the people of the region create opportunities for self-sustaining economic development and improved quality of life. ARC has worked for 55 years to bring the region into socio-economic equality with the rest of the nation.

Photo: Russel Lee. Wikimedia Commons / Public Domain.
Typical housing for coal miners. Kingston Pocahontas Coal Company, Exeter Mine, Welch, McDowell County, West Virginia, 1946. Coal transportation by rail in the background.

Milong Bond, a coal miner, is taking a bath. He rented a three-room house for $8 per month, this fee did not include the electric bill. There was no running water in the house. The mine companies owned many miners' homes and rented them out to them. Mullens Smokeless Coal Company, Mullens Mine, Harmco, Wyoming County, West Virginia, 1946.

Photo: Russel Lee. Wikimedia Commons / Public Domain.

Before closing this chapter, I want to emphasize once again that I am not writing the book to judge the people from the southeastern United States or the part of this population that has snake handling as part of their religious practice. The subject is interesting and a bit exciting and the book came about because I am a curious person. The subject is also a bit foreign to me, who only responsibly works with venomous snakes and avoids risk where possible.

My opinion is that it would have been best for the snakes and their handlers, not to be part of the religious services. The believers should demonstrate their faith in other ways. I think that many practitioners of religious snake handling probably are quite ordinary people, who during their ceremony do something that the rest of us see as quite unusual. I don't think the whole region is populated by lunatics; I don't even think the parishes are full of lunatics. However, I believe that they do not share my view on what is good or bad for snake health. And I believe that they don't share my view on un-necessary risk-taking.

Photo: Rickard Ljunggren. Two of my Broad-banded copperheads, *Agkistrodon laticinctus.*

3. Snake Church

Photo: Pixabay

Pentecostalism is a theology within Christianity originating in revival movements in Los Angeles and Wales during the first decade of the 20th century. Sometimes it is called the Pentecostal revival, the Pentecostal movement or the Pentecostal church. The name was given to the movement by the media during its emerging era in the early 20th century. The name was given in reference to the movement's emphasis on the events of the first day of Pentecost, when, according to the Acts of the Apostles, the disciples were filled with the Holy Spirit for the first time and then became bold, prophesied in different tongues and performed miracles.

Snake handling appeared in American Pentecostalism around 1910. *The Church of God with Signs Following* is a denomination founded by defectors from *The Pentecostal Church of God* after this Pentecostal movement distanced itself from the practice of handling venomous snakes during services as a sign of strong faith. Some sources claim that George Went Hensley from Grasshopper Valley in southeast Tennessee is the originator of the phenomenon. Others who have researched the subject claim that the handling of snakes during worship cannot be attributed to a single person, that the phenomenon occurred in several places at about the same time. However, historians agree that Hensley's advocacy, leadership, and especially his personal charisma, were important factors in promoting venomous snake handling and spreading it to rural churches throughout the southeastern United States. It is not known how many practitioners these congregations have today. It is partly an underground movement because they are sometimes challenged by the law and by animal rights organizations. The exact number of members is unknown, and it has recently been estimated that the total number of practitioners is as low as 1,000 - 5,000, with possibly fifty to a hundred congregations. According to the Encyclopedia of American Religions, churches *"can be found from central Florida to West Virginia and as far west as Columbus, Ohio."* The largest is *The Church of God with Signs Following* which sprang from Hensley's teachings.

Photo: Phil Dunning. Three pregnant Eastern copperhead, *Agkistrodon contortrix.* Photo taken in Clinton County, Pennsylvania.

Snake handling in the church world of North Alabama and North Georgia originated with James Miller of Sand Mountain, Alabama. Miller apparently developed his faith, during the same period, independently of knowledge of Hensley's services. While Hensley's service was trinitarian, Miller's services were non-trinitarian and are commonly known as *The Church of Lord Jesus with Signs Following*. This branch dominates snake handling churches north of the Appalachians.

Each church is independent and autonomous, and the names of the congregations are not consistent in all areas. Usually it is some variation of "*Church of God*" or "*Church of Lord Jesus*". "*Signs following*" in the names means sign followers. The five signs are found in the Gospel of Mark, which is one of the four gospels included in the New Testament.

Estimates in the last two decades have ranged from about 40 to as many as 125 congregations, all of them small, most concentrated in the Southeastern United States. While Pentecostalism has exploded globally, with tongues-speaking and faith-healing denominations growing exponentially in Latin America, Africa, and Asia, snake handling remains a practice almost entirely unique to the American South. Congregations with snake handling as part of their religious practice exist, or at any rate have existed, sporadically all the way up in Canada. There are records of four practicing congregations in Canada in the early 2000s. Most religious snake handlers are still found in Appalachia and other parts of the southeastern United States. Setting the tone are Alabama, Mississippi, Georgia, Kentucky, North Carolina, Tennessee, West Virginia and South Carolina.

Several studies have shown that the congregations are losing members, but at the same time it is the case that bans have meant that more and more venomous snake handling takes place in the dark. How many of the congregation's members actually deal with venomous snakes is impossible to answer. It can be hoped that only healthy members of legal age are allowed to come forward and drink strychnine or other poison mixed with water, dance before the congregation with a venomous snake in their hands or handle a burning object and expose their skin to the heat. In addition to poison, fire and snakes, they speak in tongues and have more speed and volume in the music than we are perhaps used to having in European churches.

Photo: Russell Lee, 1946. Wikimedia Commons / Public Domain.
Eli Sanders, pastor at the Pentecostal Church of God; he is also a tipple worker* and track loader at P V & K Coal Company, Mine, Lejunior, Harlan County, Kentucky. Most of the members, including the pastor, are coal miners with families.
* A tipple is a structure used at a mine to load the extracted coal for transport, typically into railroad hopper cars.

Reverend Eli Sanders, pictured right (with one arm), died of a snake bite during a service in September 1958. He was then 53 years old. *The Progress-Index* wrote a short article, Thu, Sep 11, 1958, page 15, (Petersburg, Virginia):

Snake-Bite Victim Not of His Church, Bishop Declares
HOPEWELL (AP) – The presiding bishop of the Pentecostal Holiness Church said today a minister who died of a rattlesnake bite at a religious meeting at Norton, Va., was not a member of his denomination. The Rev. Eli Sanders, 52, died Tuesday at his Harlan. Ky., home of the bite received Sunday in a snake-handling faith demonstration. "We have no minister by that name," said Bishop Joseph A. Synan, the general superintendent of the Pentecostal

Holiness Church. "Our denomination does not condone snake-handling in any way."

In my opinion, a bit tragic that the main focus of the article is not that a man has died prematurely, the main focus is that you give another congregation the opportunity to distance themselves from the snake handling and the accident. Regardless of what one thinks of the phenomenon, it is a person who has died and left behind a grieving family.

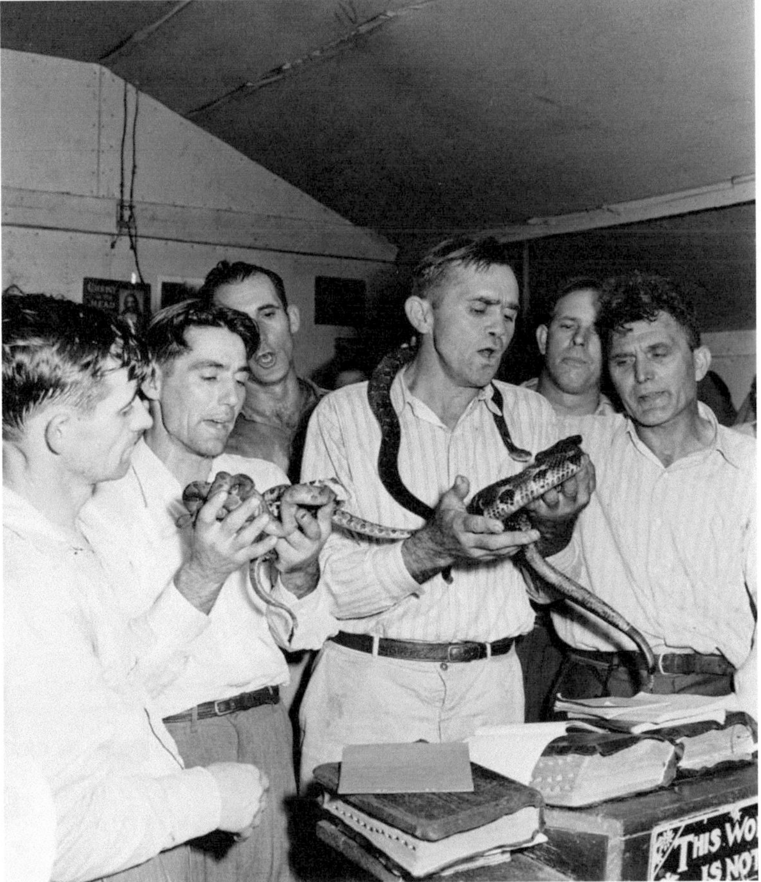

Photo: Russell Lee, 1946. Wikimedia Commons / Public Domain. The same premises as the previous picture.

Even though they only are photos, I get the impression that the room is full of chaos. Several people dance with snakes, people beat cymbals, it happened that people danced with a small torch or similar and deliberately burned themselves, they drank diluted poison. There was generally much more running and movement than we are used to seeing in churches here in Sweden.

Today there are some clips on YouTube to watch. The general impression I get is that chaos reigns and the sound level is sometimes really high with an orchestra consisting of drums, electric guitar and electric bass combined with a pastor who sometimes shouts into the microphone and perhaps someone playing the cymbals.

Photo: Steve Ludwin.
A guitar strap made of the skin from a rattlesnake. Picture taken in the House of the Lord Jesus, West Virginia.

Although snakes are not among the animals in nature with the best hearing, I believe that the noise level of these churches contributes to stress the venomous snakes that are involved.

Add to that the snakes being handled by person after person, all the people moving around in the room and all the scents these snakes take in. I believe that there is an increased risk that all of this adds up to a snake acting defensively. If you then hold the snake in your hands, it is easy for an accident to happen.

Photo: Russell Lee, 1946. Wikimedia Commons / Public Domain. (Image cropped.)

Photo: Steve Ludwin. *Crotalus horridus* in Chris Wolford's transport box. The snakes stay in the box until it is time for them to be a part of the pastor's sermon. I have seen that some parishes have terrariums or similar for the snakes to live in between services, but I do not know if this applies to all parishes and all snakes.

It is worth noting that they do not seem to try to have full control of whether the snakes will try to bite or not. I get the same picture from watching clips from modern times on YouTube. In other words, it is not about some form of herpetological skill, that one avoids bites by being able to read the snake's movement patterns. You simply don't try. Those who are bitten usually do not seek medical help but look to God for their healing.

Even though venomous snake bites and deaths have marked the history of these congregations, people continue as if they themselves were immune to bites. History repeats itself and several parishes are inherited within the pastor's family. There are families where both parish and cause of death are inherited for several generations. Various figures for the total number of deaths from snakebite during religious services have been suggested, but suggested numbers are always in the hundreds.

Religious experts agree that parishioners are aware of the risks and accept the consequences. The handling of venomous snakes carries significant risks, no matter how pure you are in your faith. Ralph Hood, who has studied these congregations, notes, "*If you go to any snake-handling church you will see people with withered hands and missing fingers. All snake-handling families have suffered such things*". Hood also notes that the handling poses no danger to those who are merely watching. There is no documented case of a church member being bitten by a snake handled by another believer.

The laying on of hands actually has an equal part in the church services, but of course does not receive as much attention from the outside world. It is the same in this book, the book's focus is on the handling of venomous snakes as that is what created curiosity in me.

Pictures next page: Church service at the Pentecostal Church of God. Lejunior, Harlan County, Kentucky, 1946. The crowd is palpable and there are plenty of children. Apparently, it rarely or never happens that someone gets in the way of a snake someone else is handling.

Photos next page: Russell Lee, 1946. Wikimedia Commons / Public Domain.

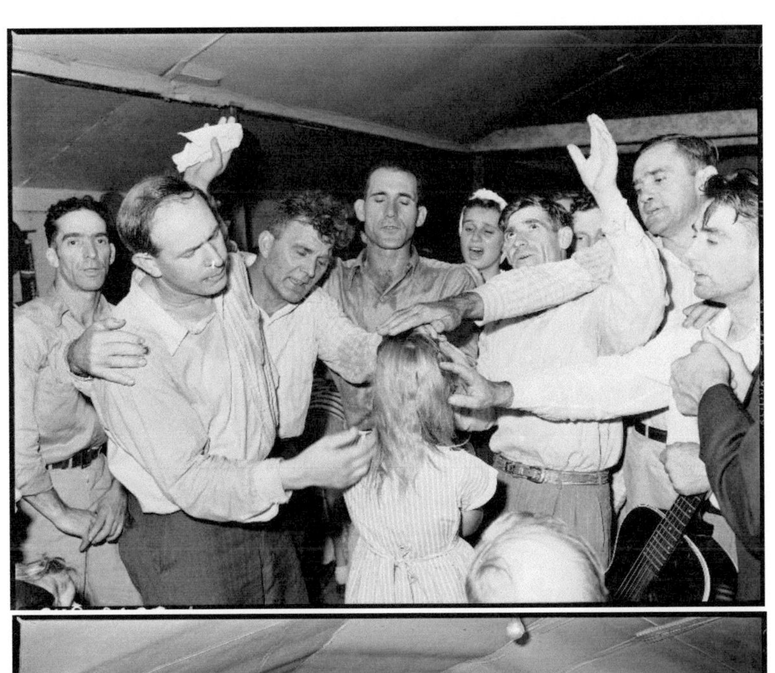

Who are the parishioners and congregations?

Snake handling is a religious rite observed in a small number of isolated churches, usually characterized as rural and part of the Holiness movement.
My picture is that they are quite small congregations, with a few tens of members in each. Many members are related to each other, a congregation may consist of a few families where the faith has been passed down for generations. The fact that the parishioners often belong to relatively few families from the surrounding area naturally contributes to the fact that it is exclusively white people you see.

The existence of close ties and a strong sense of belonging probably helps to keep a curious outside world at bay. Being the one to talk about the ceremonies or to be the one who let in the curious can perhaps be associated with the risk of being questioned within the community. This certainly closes a door that is sometimes ajar and may perhaps be the reason why someone who was initially favorably disposed has since changed their mind. More than one has stopped responding to me after initial contact and I understand that some have tried to influence others not to talk to me.
Exceptions (recruitment of new members) occur and here there is a certain over-representation of former addicts who have found God and got rid of their addiction. That a non-negligible percentage of those who find God in adulthood has a past with addiction is nothing unusual, and that is how it is in churches in Sweden as well.

If you look at who the members are, the age range is large and the gender distribution fairly even. It seems to be predominantly white people in these congregations, I have seen very few exceptions. It could be as simple as that is how the community around the parish looks like. Presumably there is some segregation in Appalachia as well. It is not entirely unusual in the United States.
Only men can become pastors, which is connected to the members' image that only a man can be the head of the family. Otherwise, men and women are equally welcome to take an active part in what happens in the church.

The members of the congregations fully believe that they must handle the snakes as a demonstration that they have the Holy Spirit within them and to show their obedience to God. Handling venomous snakes and drinking poison is for them the true test of their faith in the Bible. Even if bitten, they will often refuse to accept medical treatment because they believe that they are worthy of God's faith and that their fate is in God's hands.

Jamie Coots, a pastor who later died from a rattlesnake bite, said: "*Handlers get bitten all the time, and every few years someone dies*". Brian Pennington, a religion professor at Maryville College in Maryville, Tennessee, has studied Coots during his research on snake handling in worship. He said the prominent leader saw the practice as "a command from God." "*These are not irrational people. These are people who know very well what they are doing when they walk into the church,*" Pennington said during an interview a few years ago. "*They know very well that the fate of Reverend Coots could befall any of them who do this during a service. In many ways, I think we can expect that this type of death, especially of such a prominent figure as Reverend Coots, will cause people to recommit themselves to their faith and to actualize their obedience to a text that they feel they have a tremendous obligation to follow,*" Pennington continued. More on Coots later in the book.

Darlene Summerford, who was once married to a snake-handling pastor, was asked what it felt like to handle venomous snakes and she replied, *"It's just knowing you got power over them snakes". And, if they get bitten by the snake, then they lack the true Spirit. Moreover, if they are bitten, then the congregation prays over them".* More on Darlene later in the book. Her husband, pastor of one of these churches, tried to kill her with the help of his rattlesnakes.

*

Why don't these snakes bite more often than they actually do? It is still the case that they are constantly handled during church services, often several snakes at the same time and constantly new people who grab the snakes when it is their turn to feel closer to God.

First of all, snakes are not the killing machines that many people think. If you encounter a snake in the wild, a bite is always the snake's last resort. They try to escape, hide, threaten and strike without biting. The bite is always the last resort and snakes never chase people.
I once encountered a viper (*Vipera berus*) that coiled right over my shoe when it was about to flee to escape me and my camera. I stood in the way of escape, nothing happened because I stood still until the snake had passed.
How long the distance is before the snake feels it has no choice but to bite a human, who is not prey anyway, varies. It depends on the situation, species, individual of the snake, degree of surprise and probably also a little on the snake's daily form. Some species very rarely bite humans, but all species do so sooner or later. You can go to a church service and handle a rattlesnake or a water moccasin several times, but there is a risk that sooner or later you have handled venomous snakes that one too many times.
The snakes kept by the parishes are partly somewhat used to humans, which might reduce the risk of bites somewhat. In my opinion, the fact that the snakes are overly stressed may have a greater impact on their reluctance to bite.
On top of this, it is not unusual for snakes in poor condition to attend the service. Tests have been conducted on the snakes' feces, it is not uncommon for them to be dehydrated and malnourished and therefore have less energy to defend themselves. Some say they are not given food or water, making them lethargic and less likely to bite. Others claim the reptiles are sick and full of parasites from mismanagement in captivity. I would hope that there are also pastors who understand how to take care of their snakes, the world is usually not so simple that all people are or act in a certain way. Some handlers milk the venom from the snakes before a service to minimize the amount of venom injected in the event of a bite.
According to the believers, the only reason the snakes don't bite is because God's will is done, because the one who has the snake in his hands has the right and pure faith. Many species of snakes can often be handled if you are used to snakes, especially if the snakes are reasonably used to this happening. In the hobby of keeping snakes in terrariums, it is not entirely uncommon to handle venomous snakes with your hands.

The phenomenon is called "free handling", and it really divides the terrarium hobby into two sides that will never be able to agree. Some think it is downright insanity, unnecessary risk-taking and harmful to the reptile hobby as the risk of venomous snake bites increases significantly. It is also feared that the free handling gives the image that the hobby consists of a lot of crazies that should be stopped by banning the ownership of venomous snakes as a private person. Others handle venomous snakes themselves and claim to be able to read the snakes next move and that they know the snakes so well that the risk of a bite is as small or at least smaller than if you used a snake hook. There are also those who claim that the free handling takes place for the snake's sake, that the snakes get more nervous from feeling a snake hook than a human hand. Some who do not want or dare to deal with it themselves stand behind those who actively practice the phenomenon. On social media, several free handlers have become celebrities in the reptile hobby. Free handling seems to be a shortcut to getting lots of followers and subscribers on social media. The pictures and clips that go viral outside the hobby probably have an overrepresentation of free handlers.

My belief is that risk-taking attracts online followers, it is more exciting to see someone hold a cobra in their hands than to see a picture of someone's cobra in a well-decorated terrarium. You can compare it to the fact that a clip of someone driving a motorcycle at 200 km/h on the rear wheel probably gets better distribution than I would have gotten with a clip of me sliding forward on a moped on a dirt road in Thailand.

My opinion on free handling is that everyone who does it sooner or later has done it one too many times. So far, it seems that I am right. I do not know of any free handling influencers who have not been bitten. There may of course be a few, I don't follow this part of the hobby so well and don't know everyone who free handle venomous snakes on social media. That being said, I don't agree with those who wish for influencers to get bitten.

Photo: Phil Dunning. Timber rattle-snake (*Crotalus horridus*) at wintering site in Clinton County, PA.

My thought regarding the church phenomenon that the book is partly about is that this could be downright life-threatening. Free handling is, in my opinion, a big risk-taking both in and outside church. Those who do it in the reptile hobby probably do it most of the time under calm and orderly conditions and hopefully they are focused on the animal they have in their hands. These free handlers are also people with an interest in reptiles who have varying degrees of knowledge about snakes.

I believe that most parishioners do not have much knowledge about venomous snakes. Those who practice venomous snake handling during worship do not even seem to try to focus on the snake or in any way try to be careful to avoid making the snake nervous. It's also a little unbelievable to me that you don't learn from other people's mistakes and accidents. When others were injured or even died from handling venomous snakes, I, for one, would have asked myself if I should really continue with the snakes in the church.

If so many others have been bitten during church services, maybe I should refrain from handling rattlesnakes with my hands?

Do some people have overconfidence in their own pure faith?

Others may have been bitten and hurt, but God sees how pure I am in my faith. The snakes will not hurt me.

Or is it that for the practitioners it does not matter whether they are bitten or not. God's will be done...

Some people say that this is how they reason about the risks, which is of course very strange to me as a non-believer.

Photo: Hartwell & Hamaker. Wikimedia Commons / Public Domain. Hopi snake priest with a snake in his mouth in the Hopi Snake dance. Photo taken 1899. The Holiness movement wasn't the first to incorporate snakes into their religious rites, they weren't even the first in the US. The Hopi people are a tribe from northeastern Arizona. They belong to the cultural sphere of the *desert pueblo people* and dancing with snakes is a part of their religious practice. The most famous of the Hopi people's kachina rites is the snake dance, where they danced while handling live snakes.

4. Snakes of Appalachia

Snakes (or serpents as many in the region call them) have a central role in these churches and an even more central role in this book. My main interest is snakes, snakes have fascinated me all my life and writing about these animals is very rewarding for me. History in general has always been an interest, and I have followed the snake-handling pastors with horror-tinged fascination for many years. Religious history is far outside my scope, but the faith that this book is partly about is both spectacular and touches the snakes.

Photo: HaloJim, Pixabay

Three species of snakes from Appalachia have each received a chapter. These three species have received it as they are the species that most often are used during church services. By far the most common snake in the church world is the Timber rattlesnake, followed by the Eastern copperhead. Cottonmouths occur but seem to be rarer and venomous snakes from outside the region are sometimes also being used under service. In this chapter I will try to mention some of the species and genera that are commonly found in the area. Do not see the chapter as a complete species list, I have chosen to mention only a few snakes that I think might be interesting. The chapter also covers what the snakes in the area have in common.

Crotalus horridus – Timber rattlesnake.

The Latin word *Crotalus* derive from the Ancient Greek *krotalon*. The latter term referred to a type of clapper, rattle, or castanet as used in ancient Greece and Egypt. *Horridus* means "horrible/frightful".
This is the only rattlesnake species in large parts of the densely populated northeastern United States. The species is second only to the Prairie rattlesnake, *Crotalus viridis*, the most northerly distributed venomous snake in North America. *Don't Tread on Me* is a political slogan that dates back to the American Revolution and is still widely used today. The Gadsden flag is a historic American flag with a yellow field depicting a rattlesnake coiled and ready to strike. Below the rattlesnake are the words *Don't Tread on Me* and the symbolism is crystal clear, step on me and I will strike back! The flag is named after politician Christopher Gadsden (1724–1805), who designed it in 1775 during the American Revolution. In modern times, it is sometimes used in the United States as a symbol of conservatism, generally rebellious sentiments, and sometimes of far-right ideology.
Metallica released the song Don't tread on me in 1991. Singer James Hetfield describes the song as *"just one of those don't fuck with us songs"*. I would think that is a good summary of the symbolism behind both the flag and the lyrics as that it is a good summary of the Timber rattlesnake as well. You can read more about the species in chapter 8.

Photo: Wikimedia Commons / Public Domain

Crotalus adamanteus – Eastern diamondback rattlesnake.

This rattlesnake is the largest rattlesnake species and one of the heaviest known species of venomous snake in the world. A specimen shot in 1946 measured 2.4 meters in length and weighed 15.4 kg! [6] Other venomous snakes can rival this species in weight, such as the much longer but slimmer King cobra (*Ophiophagus hannah*) of Asia and the shorter but even stockier Gaboon viper *(Bitis gabonica)* of Africa. Usually, snakes of this species do not grow this large, anything above 170cm in length is regarded as very large. The average size is much smaller at 130cm. Males of this species are usually larger than females.

The species occurs in the southeastern United States in the coastal states from North Carolina down to Florida and west to Louisiana. Like most rattlesnakes, this species is terrestrial and not particularly a skillful climber. However, the species has occasionally been reported in bushes and trees in search of prey. Even large specimens have been discovered as high as 10 m up in the branches of bushes and trees. The Eastern diamondback rattlesnake is reputed to be the most dangerous venomous snake in North America. The species is not aggressive, just like other snakes it should be considered defensive. Snakes bite to kill prey, or to defend themselves.

It is the species responsible for the most fatal bites in the United States and studies show a 10-20% fatality rate for untreated bites [7]. Effective serum exists, but it is not unusual for a patient to need many doses in severe cases of envenomation. One popular myth is that the eastern diamondback rattlesnake must rattle before striking. On the contrary, it is quite capable of striking while remaining completely silent.

Like some other snake species in the region, this rattlesnake has several common names, but Eastern diamondback rattlesnake is the most common. I am not going to list these other names in the text as I think it is a good thing if we stick to one common name to avoid confusion. It is of course best to use scientific names, but everybody is not interested in learning these.

Photo: Dr. Edward J. Wozniak. *Crotalus adamanteus. Adamanteus* means "diamond" in Latin.

Photo: Rickard Ljunggren. A *Sistrurus miliarius (barbouri).*

Sistrurus miliarius - Pygmy rattlesnake. The subspecies *barbouri* on the previous page is called (among other names) Dusky pygmy rattlesnake in English.
As the name suggests, it is a small species, often 40-60 cm long. The small rattle emits a buzzing sound that can only be heard a few meters away. Since the species cannot produce much venom, it is unlikely that it will be able to deliver a fatal bite to an adult human. Children who have been bitten have sometimes had a prolonged hospital stay, with reports of tissue death around the bite.
The species occurs throughout the southeastern United States and further into Texas and Oklahoma. The species inhabits flatwoods, sandhills, mixed forests and floodplains, and is also found near lakes and marshes. Due to the need for cover they may prefer more densely vegetated areas. (Flatwood is a soil series with impaired drainage that occurs in the southeastern United States.) Just like other rattlesnakes, *Sistrurus miliarius* does not lay eggs.
The species is opportunistic in its diet, which can be said about almost all snakes. The diet includes small mammals and birds, lizards, insects, desert centipedes and frogs, as well as other snakes. Some prey, like the centipedes, are hunted by active pursuit, grabbing and flipping them around while injecting venom to prevent injury.
More common is the ambush method when the snake waits for the prey to come to close. Young specimens can use caudal luring when hunting skinks, but this method becomes less effective for adults, as the prey size and type changes. Their feeding strategy becomes sit-and-wait, with individuals remaining in a coiled position for several days at a time.
Caudal luring is a form of mimicry characterized by the predator's tail wagging or wriggling to attract prey. This movement attracts small animals that mistake the tail for a small worm or some other small animal. When the animal approaches the "prey" (the worm-like tail), the predator will strike. The phenomenon is found in many viper species. These snakes often have a tail that differs in color from the rest of the body, and in several of the species the color difference disappears when they become adults. Think of the bright tail as a hook on the end of a fishing line!
The name of the genus *Sistrurus* comes from the Latin word *sistrum* meaning "a rattle". The specific name, *miliarius*, is derived from Latin

and means millet or millet-like, perhaps referring to the blotched patterns described in the species.
It is a species that is not uncommon to have in a terrarium, thanks to its beautiful appearance and small size.

Agkistrodon contortrix – Eastern copperhead.

There are an estimated 7,000 – 8,000 venomous snake bites per year in the United States, but fewer than ten people per year die after being bitten. The same statistics show that 10 - 44% of those who have been bitten by a venomous snake suffer permanent damage [8], it is not like in movies where the hero ingests the antidote and immediately becomes fully healthy again. In the southeastern United States, the Eastern copperhead is responsible for the majority of venomous snake bites. The explanation is that the species is quite common and very well camouflaged. In a pile of leaves it is very difficult to detect a copperhead lying still. Fortunately, the Eastern copperhead's venom is not as potent as some other medium sized vipers and fatalities are extremely rare. Most snakebites occur when someone tries to kill or disturb a snake, so the best way to avoid a bite is to leave a snake you find alone. The same can be said about all venomous snakes, many who are bitten are bitten because they do not leave the snake alone. Myths about venomous snakes hunting people exist in the United States, just as in other parts of the world. You can read more about the species in chapter 9.

Photo: Dr. Edward J. Wozniak. Eastern copperhead, *Agkistrodon contortrix,* in Little Rock, Arkansas.

Agkistrodon piscivorus – Northern cottonmouth, Water moccasin, Swamp moccasin among others.

This species is known to open its white mouth when it feels threatened. A good way to warn those who come too close if you don't have a rattle! The phenomenon is found in other species in the world, the most famous is probably the Black mamba (*Dendroaspis polylepis*) in Africa. The Black mamba itself is not black as you might think, the Black mamba gapes and shows its black mouth so that we can identify it as deadly and leave the snake alone.
You can read more about the species in chapter 10.

Nerodia - water snakes.

Nerodia is a genus of snakes that live their lives in or near water. They are quite roughly built and often have a color scale similar to that of cottonmouths. It is not uncommon for snakes in the genus *Nerodia* to be confused with these and even the other way round. People touch a cottonmouth in the belief that it is a harmless snake that often bites but has no venom. Or they kill a snake because, in ignorance, they see it as a threat. Of course, you shouldn't kill cottonmouths either, you should only kill wild animals if there is a danger to your own health or if it is a matter of hunting.
The first instinct of these snakes is to flee when disturbed, but they will defend themselves if they cannot escape. They will not hesitate to bite if handled, often multiple bites. They can also excrete a foul-smelling musk from their cloaca, just like snakes in the European genus *Natrix* and some other colubrids can. Species of *Nerodia* are widely distributed throughout the eastern half of the United States. They live in swamps and near bodies of water, are good swimmers and hunt amphibians and fish. Females do not lay eggs but give birth to live young. In individual specimens, the number of young per litter can be up to 100 [9]!

Photo: Dr. Edward J. Wozniak. Northern cottonmouth, *Agkistrodon piscivorus,* displaying its famous white mouth.

Photo: Washu, Pixabay. A Common watersnake, *Nerodia sipedon.* The species is frequently mistaken for venomous cottonmouths, *Agkistrodon sp.*

Thamnophis - garter snakes.

Garter snakes are closely related to the genus *Nerodia,* with some species having moved back and forth between the genera. Garter snakes are harmless, small to medium-sized snakes belonging to the genus *Thamnophis.* The many different species are found across most of North and Central America, from Canada down to Costa Rica. They are fairly small snakes, ranging in length from just under half a meter to over a meter.

Garter snakes have a system of communication through body odor, they can find other garter snakes by following their pheromone-scented trail. Male and female skin pheromones are so different that they are immediately recognizable for other garter snakes, but male garter snakes sometimes produce both male and female pheromones. During the mating season, this ability tricks other males into trying to mate with them. This causes a transfer of body heat from the male being fooled, to the other snake, which is a great advantage during mating season. The male who is getting warmed will be more alert and active and will find it easier to compete with the other males. Male snakes that emit pheromones from both males and females have been shown to have more matings than other males.

Garter snakes were long thought to be non-venomous, but it was discovered in the early 2000s that they actually produce a neurotoxic venom. However, garter snakes cannot harm or kill humans with the small amounts of relatively mild venom they produce. In a few cases, some swelling and bruising have been reported.

Garter snakes are not uncommon to keep as pets in a terrarium, I have had several different species myself. A garter was the first snake I bought as soon as I moved away from my parents' home and the choice fell on a *Thamnophis sirtalis.* A small and easy-to-keep species that was much cheaper than the Corn snake (*Pantherophis guttatus*) I really wanted. This was in the mid-90s and the range of available species that exists today was not available in Sweden at that time.

Photo: Travis Sova. Eastern garter snake, *Thamnophis sirtalis sirtalis,* in Upstate New York.

Photo: onkelramirez1, Pixabay. Two Corn snakes, *Pantherophis guttatus*.

Pantherophis guttatus - Corn snake.

The Corn snake is named after the regular presence of the species near grain stores, where it hunts mice and rats that eat harvested corn. According to the Oxford English Dictionary, the name has been used since at least 1675. Other sources claim that the corn snake is so named because the distinctive, almost checkered pattern on the snake's abdominal scales resembles the kernels of variegated maize. If you like tall tales, I read somewhere that the first European settlers gave the species its name because they were thought to eat corn, as these snakes were often found near cornfields.
The corn snake is probably the most common snake to keep as a pet and, in my opinion, a species that is very easy to manage.

Pantherophis obsoletus - Western rat snake, Black rat snake, Pilot black snake, or simply Black snake.

Photo: Daina Krumins, Pixabay. Two *Pantherophis obsoletus*.
These are long, slender, harmless snakes that constrict their prey, just like the Corn snakes that belong to the same genus. As their name suggests, these snakes are very dark in color. Snakes from these two

Pantherophis species can grow to over 150cm long and both are often found near human settlements. Both species are of great benefit to people in the region as they eat large numbers of rats, mice and other pests. For this reason, farmers appreciate having these snakes around. When black rat snakes or corn snakes get into a tight spot, they will often coil up in a defensive position, hiss and lunge. They also vibrate their tails rapidly, like a rattlesnake even though they have no rattle. They simply borrow the rattlesnake's mimicry to fool the outside world. During winter the black rat snake hibernates in shared dens, often with Eastern copperheads and Timber rattlesnakes. This association gave rise to one of its common names, Pilot black snake, and the superstition that this nonvenomous species led the venomous ones to the place for hibernation [10].

Lampropeltis getula - Common kingsnake, Eastern kingsnake or Chain kingsnake.

This nonvenomous snake is a species with about ten subspecies. Kingsnakes are not as long as the snakes in the genus *Pantherophis* on the previous page, they are usually just under a meter long. The species is found across much of the United States and, for some years now, in Gran Canaria. In Gran Canaria, they are released pets that thrive in their new environment and have become an invasive species. The island used to have no snakes, but now there are so many that it has become a problem for the native animals. As a result, these kingsnakes (all subspecies) are now banned from being bought or sold in the EU. Kingsnakes eat other snakes, including venomous snakes such as Eastern copperheads, *Agkistrodon contortrix*, which are responsible for more venomous snake bites than any other species in the United States. Kingsnakes have evolved a hunting technique to avoid getting bitten, clamping the venomous snake's jaws with their mouth during the fight. Should the kingsnake be bitten anyway, the venom has no direct impact on the snake's health. They also eat amphibians, turtle eggs, lizards and small mammals, the prey are being constricted until death.

Coluber constrictor - Eastern racer, Black racer or North American racer.

These snakes are dark in color, as one of the common names suggests, and up to 150cm long. Juveniles are more strikingly patterned, with a middorsal row of dark blotches on a light ground color. As they grow older, this snake darkens, and the juvenile pattern gradually disappears. These racers are very fast and typically flee from a potential predator. However, if cornered, they will put up a fight, biting repeatedly. They are difficult to handle and will writhe, defecate, and release a foul-smelling musk from their cloacae. Just like so many other nonvenomous colubrids they can vibrate their tails among dry leaves, and sound convincingly like rattlesnakes.
"Runner" is sometimes used instead of "racer" in their common names.

All snake species in the region hibernate during the winter, with many species hibernating together. They find shelter in hollows in the ground, caves or cellars and house foundations. There they spend the winter in an inactive state similar to hibernation. Timber rattlesnakes and Eastern copperheads prefer south-facing rocky outcrops. These need deep crevices to hibernate and ledges to bask on during late autumn and early spring. Biologists claim that many of the hibernation sites have been used for centuries and some may even be ancient. Doesn't sound unreasonable to me, as long as we humans don't destroy nature, the snakes return to the same hibernation site every autumn. Here at home in Sweden, I visit the same hibernation site every spring and probably see the same vipers every spring as long as they survive another year. Some species sometimes share the same den, usually it is *Coluber constrictor, Crotalus horridus* and *Agkistrodon contortrix* that can be found together.

Photo: Phil Dunning. Crotalus horridus and Agkistrodon contortrix hibernating together.

5. A collection of characters

This chapter presents a small selection of interesting people within the part of the Pentecostal Church that the book tries to tell a story about. I have probably missed several people who should have been included in the book, so you can see the chapter's cast of characters as a selection. Some of the people who have been included in the book have been significant to the practice of faith that this book is about; others have been included for the simple reason that I see those particular people or their story interesting.
Some people in this world have taken an active part in the media. They have appeared in programs on National Geographic or other television production companies, or they have appeared on YouTube and on social media. Some have willingly given interviews, but that time seems to be over.

Getting in touch with these people was not easy. I found some things written about them, but they tend to avoid my attempts to start a dialogue. In general, most people nowadays avoid attention as snake handling is not completely legal everywhere and previous attention has usually been negative and judgmental. Practitioners have been declared idiots or prosecuted, had snakes confiscated and killed. In addition to the legal problems that snake handling can cause, churches have had problems with animal rights organizations and with congregations that do not use venomous snakes, poison or fire as part of their worship.

Practitioners claim that their constitutional rights are being violated when the law and the state try to prevent congregations from using snakes during worship. The issue has never been tested in higher court, but there are those who believe the practitioners could be vindicated. The human right to potentially subject snakes taken from the wild to unnecessary stress for religious reasons is likely to be prioritized over what is considered to be in the best interests of animal welfare.

George Went Hensley

It seems that everything started with him, so it is natural that this chapter also starts with him. He wasn't the first to do this, but he was the first to get well known outside the Pentecostal snake handling movement.
George Went Hensley (2 May 1881 - 25 July 1955) was an American Pentecostal minister who is best known for popularizing the practice of venomous snake handling during church services. Born in rural Appalachia, Hensley experienced a religious conversion around 1910, when he interpreted the Bible's New Testament to command all Christians to handle venomous snakes.

According to an old legend, Hensley wandered into the wilderness to seek God's will, which he was pretty sure would involve snakes. While walking over the hills, he came across a snake, knelt over it in prayer and picked it up. It didn't bite him, and he chose to take the snake with him. He then took the snake to his congregation, whose members were naturally impressed and decided that snake handling was a good idea. Then, depending on whether you ask Hensley's followers or historians, spiritual awakening either broke out or not. But whether or not many souls were saved, many snakes were definitely handled.
After his conversion, he travelled through the southeastern United States teaching his faith, which emphasized strict personal holiness and frequent contact with venomous snakes. Although illiterate, he became a licensed minister in *the Church of God* (Cleveland, Tennessee) in 1915. After travelling through Tennessee for several years conducting services approved by *the Church of God*, he resigned from the denomination in 1922. Many years later, Hensley began calling his church *the Dolly Pond Church of God with Signs Following.* Today, the churches that have taken over Hensley's legacy are called *Church of God with Signs Following.*

Hensley was married four times and had thirteen children. He had many conflicts with his family members due to drunkenness, frequent travelling and inability to have a steady income. At least, these reasons were given by his first three wives for their divorces. Hensley was arrested for moonshining in Tennessee during Prohibition and sentenced to community service. However, he chose to flee and evade justice. Hensley travelled to Ohio, where he held revival meetings in several different locations. His services ranged from small meetings in someone's home to large gatherings that attracted media attention and hundreds of participants. Despite holding many services, he did not make much money, and he was arrested for violating snake handling laws at least twice. Hensley's theology, with the exception of venomous snake handling, was typical of other fundamentalist Pentecostal churches. In his sermons, he condemned many things as sinful, including gambling, drinking alcohol, wearing lipstick and playing baseball. Condemning the consumption of alcohol may seem a bit rich, coming from an old moonshiner whose own alcohol consumption had led to more than one divorce.

The 17th and 18th verses of chapter 16 of Mark's Gospel, the "longer ending" of disputed authenticity, formed the core of Hensley's justification of snake handling and the ingestion of poison during worship. Although venomous snake handling was the most common practice, he also drank strychnine or battery acid. He interpreted the passage as a commandment, rather than an observation of events that occurred in the lives of certain apostles as Christians have traditionally interpreted the verses.

And these signs shall follow them that believe; In my name shall they cast out devils; they shall speak with new tongues; They shall take up serpents; and if they drink any deadly thing, it shall not hurt them; they shall lay hands on the sick, and they shall recover.

The same verses are the basis for the religious practice of the other snake-handling churches.

By handling snakes, Hensley saw himself as part of an ongoing tradition that originated in a New Testament injunction. He maintained that the ability to handle venomous snakes without harm as evidence of salvation and steadfast faith. For him, snakes represented the devil and snake handling was an affirmation of God's power to save people from harm. He interpreted the legal difficulties he encountered as religious persecution and those who rejected snake handling were, in his view, lacking faith.

Hensley claimed to have been bitten by many venomous snakes with no ill effect and by the end of his career he estimated that he had survived more than 400 bites. If you ask me, it sounds highly unlikely that someone would have been bitten 400 times by venomous snakes without learning a lesson. It is even more unbelievable that someone could survive 400 venomous snake bites without seeking medical care.

In 1955, while conducting church service in Florida, he was bitten by a snake and became violently ill. He had a series of revival meetings scheduled near Altha, Florida, when it happened. He led the meetings without snakes for three weeks, before acquiring a rattlesnake that was reportedly 1.5 meters long. He took this newly acquired rattlesnake to a Sunday afternoon service on 24 July, in an abandoned blacksmith shop where a crowd had gathered. During the service, Hensley delivered a loud sermon on the subject of faith. He took the snake out of the jar it was kept in, wrapped it around his neck and rubbed it on his face. He walked round the audience while preaching and would then put the snake back in the jar.

When he put the snake in the jar, it bit him on the wrist. After a few minutes, it was obvious that Hensley had reacted badly to the snake's venom. He experienced severe pain, a discolored arm and bloody sputum. Despite this, he refused to receive medical attention. An eyewitness claimed that Hensley blamed his suffering on the congregation's lack of faith. He died early the next morning. Calhoun County Judge Hannah Gaskin listed suicide as the cause of death.

Hensley was buried two days after his death in a cemetery 3.2 kilometers from the forge where he was bitten. After the funeral, representatives from some of the parishes met and declared their intention to continue handling snakes. Hensley's fourth wife Sally decided to continue spreading her late husband's teachings and said after the event that she had not lost "an ounce of faith".

Hensley had convinced many rural Appalachians that venomous snake handling was commanded by God and his followers continued to do so after his death. Although the custom of handling venomous snakes developed independently in several places, Hensley is generally credited with spreading the custom in the southeastern United States.

James Miller

According to many sources, the book's phenomenon was inspired and pioneered by George Went Hensley. James Miller is a contemporary of Hensley but has not been credited by posterity with the same influence. James Miller was from a Pentecostal background, which places a strong emphasis on experiential faith, healing, and the supernatural. Like many practitioners in this tradition, he believed in demonstrating his faith through signs, including snake handling. Miller participated in snake handling as a part of worship, viewing it as a demonstration of faith in God's power to protect believers from harm. He participated in church services where venomous snakes were handled, believing that such actions were a test of faith and obedience to God. As a pastor, Miller's approach influenced his congregation and those who shared similar beliefs. He often preached about the importance of faith and living by the Word of God, encouraging his followers to embrace the more controversial aspects of their faith.

Religious snake handlers in Alabama and Georgia claim a legacy from both James and George. James was a preacher who independently began the practice in 1912 in Sand Mountain, Alabama. Under his influence, the movement spread to Georgia in the 1920s. *The Church of God with Signs Following* is run by followers of Miller's teachings.

While James Miller is not as widely known as some other figures in the snake handling community, his belief in handling snakes as a test of faith highlights the broader culture of religious fervor where adherents might take extreme risks. Information about the specifics of his life and untimely death may not be as widely reported as that of others in the snake-handling tradition. James Miller's beliefs reflect a unique intersection of deep faith, cultural tradition, and the often risky behaviors that come from interpreting scripture in a literal way.

The Brown family

A 28-year-old woman in Jamie Coots' congregation, named Melinda Brown, was bitten on the arm by a Timber rattlesnake in 1995 during a church service that Coots was leading. She died two days later at Jamies home, from the effects of the bite. The mother-of-five's relatives disputed statements that she herself had held the rattlesnake that bit her, and they also disputed claims that she had refused medical treatment. The Coots were charged in connection with Melinda's death, but a judge decided not to pursue the case.

Melinda's husband was John Wayne "Punkin" Brown, who used to accompany Jamie Coots on his tours to other churches. According to him, "Punkin" had survived 22 venomous snake bites. John Wayne "Punkin" Brown was granted sole custody of the five children, but custody had been granted on two conditions. The conditions were that he agreed not to have venomous snakes in his house and that the children would not be allowed to attend snake-handling services. He defied these orders, sincerely believing he was doing God's will, even though the children were known to wake up screaming from terrifying nightmares about snakes.
Evangelist "Punkin" Brown followed in his wife's footsteps during a 1998 sermon in Alabama.
"They say it won't bite!" Brown roared as the rattlesnake writhed in his grip.
"If it won't bite, there's no point in being afraid!
The Lord told me it was okay!
The Lord said it would be all right!"
Then, when the preacher jumped on stage, what has happened to many others happened to him. The rattlesnake bit him on the left middle finger. He continued as if nothing had happened, and the congregation did at first not realize that he had been bitten.
"God is still God, no matter what comes." He said in a voice that had lost all its power.
A woman in the congregation screamed and other members anxiously wiped the dying preacher's forehead.
"No matter what else, God is still God." were his last words, according to witnesses.

Ten minutes later, he was dead, and his five children had been orphaned. The five children then became pawns in a custody battle between their grandparents. In court, religious tradition was pitted against childcare laws, faith against science.
Their grandmother on the mother's side wanted to keep them as far away from snakes as possible as she herself had left religious snake handling behind, their grandparents on the father's side ran their own snake handling church. Custody fell to the children's grandmother and the children avoided a future with the snakes they feared and had nightmares about. "Punkin" himself had previously said in an interview that the eldest son was terrified of snakes after witnessing the bite that killed his mother.
"Jonathan is terrified of snakes!" he said but added that he had done what he could to prepare the children.
"The children knew what to expect if the Lord did not intervene." he continued during the interview.
"I told the children that Melinda had died, and they have said nothing to me to suggest that they held me responsible for her death."
In 1995, when the children's mother was bitten and died, Jonathan was nine years old. He was the oldest of the children and none of the children were, in my opinion, old enough to discuss responsibility or in any way hold anyone responsible in a discussion with their father.
In my opinion, all three quotes from the interview are examples of statements by a religious fanatic. The interview was conducted by Scott W. Schwartz, the author of the book *"Faith, Serpents and Fire: Images of Kentucky Holiness Believers".*

Ralph Hood, a psychology professor at the University of Tennessee-Chattanooga and friend of the Browns, said during the time that snake handlers forbid their children to touch serpents and often keep the kids at the back of the church. He also said that he had never heard of a child being killed or even injured at a snake-handling service. Removing the Brown children from a loving home and from the church their parents died for is not in their best interests, according to him.

Glenn Summerford

If Glenn is not released sooner, he will remain in prison until year 2121, or at least until his own death.
In 1992, snake handler and Pentecostal pastor Glenn Summerford was sentenced to 99 years in prison for the attempted murder of his wife, Darlene Summerford. For those of you who were just counting on your fingers, I can add the 30 years that Glenn received in additional sentence after a prison escape that did not last very long.
On 4 October 1991, a violent crime was reported in the sleepy town of Scottsboro, Alabama. An ambulance was called to a dilapidated house on the outskirts of town. The driver of the ambulance was asked to turn off the headlights and quietly approach the residence. Darlene Summerford was waiting for them in the driveway. She was holding a bleeding wrist. The skin surrounding the wound had already begun to blacken, and the hand had begun to deteriorate from tissue damage.
According to her testimony, Glenn had been drunk, jealous and aggressive when he grabbed her by the hair and dragged her to a shed at the back of the property. In the shed, Glenn kept about fifteen venomous snakes that he used during his sermons.
At gunpoint, he forced her hand into a box of rattlesnakes. The plan was that she would die from a snake bite and that he would thereby be innocent of her death.
Darlene testified that he wanted to marry another woman, so according to her, he felt it was easiest for him if she died first.

"He took an iron pipe and hit the boxes hard, so that the snakes got really angry, and then grabbed me by the hair and said that he would push my face into the box if I didn't put my hand in voluntarily,'" she said in court.
"He said I had to die because he wanted to marry another woman."
Darlene had two snake bites on her hand. Glenn refused to let her seek medical attention. She finally managed to call 911 and get an ambulance to quietly pick her up before it was too late.
Darlene was taken to the local hospital and then rushed to another hospital that had serum.

When she recovered, Darlene pointed the finger at her husband for what had happened. The gruesome story and the subsequent trial received a lot of media attention. Glenn Summerford was convicted of attempted murder. Two prior felonies and a lengthy criminal record nearly handed a death sentence to the disgraced pastor, who had lived a life of violence, drinking, and trouble with the law before finding God.

"Salvation on Sand Mountain" is a 1995 book by Dennis Covington. The storyline follows the author as he goes from covering the trial of Glenn Summerford to experiencing a snake handling church in Appalachia. The book, which is written in the first person, begins in a neutral, journalistic style and becomes more emotional as the author is drawn to the people and practices of the church.

The crime and Glenn's tumultuous life are also depicted in the HBO documentary *Alabama Snake*, which premiered in 2020. In my opinion, it's a fairly watchable film, not as sensationalist and dramatic as many American documentaries about murderers or venomous snakes can be. Alabama Snake mixes old VHS footage of their church services, police videos of their home and interviews with the family and people close to them. The film underscores the decaying environment of Glenn and Darlene's life together and the unbeatable fire-sulfur-booze environment they called their own.

Not only Darlene, but also Glenn's first wife Doris appears in the documentary.

"I loved him but sometimes I was afraid of him. Fear and love, they went together," Doris says in *Alabama Snake*. She recalls how Glenn almost murdered a man during an organized fight over money.

"I knocked his eyes right out of his head, I hit him on the side of the head and his eyeball came out of his nose," Glenn himself said on a tape recording for the film.

In retaliation, Glenn's house was set on fire, and the fire killed his youngest daughter.

"He was never the same after that," said Doris.

The film depicts Glenn's life, from his early years up to and past the attempted murder of Darlene. Glenn learnt to fight at a young age, his stepfather taught him. He turned out to have a talent for violence, if you can call it that. Glenn grew up to be a young man whose taste for alcohol went hand in hand with his penchant for violence.

After finding God, booze and aggression remained in Glenn Summerford's life. It is not a pretty picture we get in *Alabama Snake*. The film tells a story about a horrible person who lived a horrible life in a not-so-appealing environment. See the film!

This is not the first film Glenn and Darlene have appeared in. Back in 1991, they were in the film *In Jesus name* which is a documentary about a snake-handling pastor who was arrested, and the film depicts the legal aftermath.

The accusation of Darlene's attempted murder shook the radical Christian community. What followed was a sensational and liquor-saturated tale of adultery, religion, and attempted murder by rattlesnake that continues to haunt Southern Appalachia nearly three decades later. Glenn claimed that Darlene had been unfaithful with another preacher and that that was his reason for taking her to the snakes (and thus closer to God). A defense witness told jurors that Darlene Summerford told her that *"she got Glenn so drunk he passed out and she went out to the shed to get a snake to put on him but it bit her instead".* At the time of the verdict, his congregation claimed he was innocent. In their eyes, his wife had been evil, and the snake had good reason to attack her.

In 2004, the book *The Serpent And The Spirit: Glenn Summerford's Story* was published. This book tells the story of Glenn's rise and fall through interviews, court documents and other sources. Free of any prejudice against or romanticization of snake-handling religion, the book presents a story of a fascinating group of people, while allowing the reader to draw their own conclusions about Glenn's guilt or innocence.

The extra 30 years we got from counting on our fingers? In 2003, at the age of 58, Glenn Summerford escaped from prison. 45 minutes later, he was caught and brought back to prison. It might seem like a bad deal to trade 45 minutes of freedom for 30 years in prison, but for Glenn, it doesn't matter. He is approaching 80 years old and is unlikely to even survive his 99-year sentence for the attempted murder.

He is being held at the Bullock County Correctional Facility in Alabama. He was denied parole in 2020, 2021 and 2024. Today he is 79 years old, he will probably die in prison. Glenns cousin Billy Summerford, pastor of *Old Rock House Holiness Church* in Jackson County, carries on the tradition of snake handling today.

The Wolford family

Chris Wolford

In September 2024, I contacted Chris Wolford, the pastor who, in the clip I had seen a couple of years earlier, was standing on a rattlesnake while preaching. I told him about myself, my previous book on the subject and my current project. I wrote that I had gotten two pastors to tell their stories and opinions and that I was contacting him because I was looking for a third. Since many people are suspicious, I added that the interviewee gets to read the text before it goes to print, so that everything is correct, and they do not feel deceived. I also added that my opinions were not going to shine through too much, the reader should form their own opinion on the subject.

Photo: Steve Ludwin. Chris Wolford preaches and handles a *Crotalus horridus* in his church in West Virginia. The congregation member to the left is also handling a Timber rattlesnake.

The answer is short and not what I was hoping for.

"I don't care to do a interview with you do you mind telling me who is the two ppl you have spoken with in the faith"

There will simply have to be a few lines about Chris Wolford without his participation. Chris is the pastor of the *House of the Lord Jesus* in West Virginia, in what several sources claim is the last church in the state whose services focus on the five signs*. He is relatively open about the way he practices his faith. In addition to handling venomous snakes during church services, Chris drinks diluted strychnine, holds a flame to his skin and heals parishioners. He is quite open about his adherence to the five signs, but not so open about the snakes having a place in the church. Chris does not advertise that his church practices serpent handling on the outside, after receiving threats to burn down the building from rival Christian groups who believe the practice is against the word of God. At the same time, he appears quite often in articles, often with pictures that he sometimes contributes himself. And since snakes are what arouse the curiosity of the outside world, that's what the articles often focus on and what is visible in the pictures.

* *"These signs shall follow those that believe: In my name shall they cast out demons; they shall speak with new tongues; they shall take up serpents; and if they drink any deadly thing, it shall not hurt them; they shall lay hands on the sick, and they shall recover."*
Jesus, in Mark 16:17-18

Chris keeps around half a dozen fully grown snakes in the basement underneath his church. Usually, the snakes are caught from the surrounding countryside, and kept with heat lamps in glass cages and they are fed mice. A picture of Jesus watches over the snake den. Chris rejects any claims stating that he mistreats the snakes - and has passed inspections from local authorities. I have seen pictures of the animal housing. They are perhaps not exactly aesthetically pleasing terrariums, but more like aquariums with a heat lamp on top, bottom substrate, a plastic box as hiding place and a water bowl. In other words, no worse than how some in the reptile hobby keep their snakes. Chris has said on more than one occasion that he is afraid of snakes, but that he doesn't feel the fear when he handles snakes in the house of God. According to him, he has been bitten five times, three times by rattlesnakes and twice by copperheads.

Chris's family has practiced their religion with venomous snakes and drinking poison for generations. This seems to be a recurring pattern, many practitioners have grown up in this environment and all or most of their family have belonged to a church where snakes, fire and poison are part of the worship service. There seem to be very few, if any, pastors who are not born into this more unusual part of the Pentecostal movement.
In 1983, when Chris was 11 years old, his father died after a snake bite that did not get medical attention. Mack Ray Wolford died a slow and painful death as the snake's venom coursed round his body stopping his blood from clotting, causing him to hemorrhage to death and taking his last breath about eight hours after the snake bite. Mack Ray Wolford was not a pastor; he was a congregation member. Chris's uncle Brady Dawson was the pastor in the congregation.
Brady was interviewed about the accident.

"Wolford, who previously had been bitten three times, refused medical treatment", said Brady Dawson, minister of the church and Mack Ray Wolford's brother-in-law.
"As he lay dying on a mattress before the altar, members of the congregation and others from a nearby church prayed over him and sang, accompanied by electric guitars and drums. Two or three minutes before he died, he was laughing, joyful. "He was 100 percent God."
"It is essential to believe the fullness of the Bible," said Dawson and explained that the ritual dates back to the fourth century and is designed to show obedience to God. *"People say we murder ourselves, but we martyr ourselves."*
The minister said his church rarely practices strychnine drinking. *'When you tell people why you want it, they don't want to sell it."*

Chris's brother "Mack" Wolford also died (2012) from the dangerous practices with snakes, but despite everything, Chris continues practicing snake handling. Both his father and brother chose not to go to hospital, they were willing to give their lives for what they believed in.

Chris Wolford decided to open his own church after the death of his brother, who was also a pastor in the faith. Chris, a recovered drug addict, set up the church after he said: "God saved him" from addictions to crystal meth, cocaine and pain pills - and his congregation grown from just three to over 30 in just two years. Many members of the congregation claim that they have been healed from other addictions such as alcoholism as well as illnesses such as lupus and heart conditions. I have the impression that his parish has a high proportion of former addicts.

*

He made headlines in August 2019 when he was bitten by an eastern diamondback* held by a parishioner during a church service:

Snake-handling preacher left unconscious, paralyzed and suffering kidney failure after near-fatal snake bite at church – but vows not to stop dangerous practice

* The eastern diamondback, *Crotalus adamanteus*, is one of the heaviest venomous snakes in the Americas and the largest rattlesnake. The species has the reputation of being the most dangerous venomous snake in North America. While not usually aggressive, it is large and powerful. Mortality rate of 30% have been mentioned, but other studies show a mortality rate of 10–20% in untreated bites.

Within 30 minutes after the bite, Chris became paralyzed, his lips and tongue began to swell, and he started to struggle to breathe. He wanted to avoid hospital, as he believes his life should be in God's hands. In the end he was rushed to the ER by concerned followers after he fell unconscious.

At hospital, his kidneys stopped working as his body struggled to cope with the snake's venom coursing round his veins, and he begged doctors to amputate his arm he was in so much pain. Initially Chris refused to let the doctors start dialysis treatment, he waited as long as possible, praying for a miraculous recovery. Finally, just as the doctors started dialysis, his kidneys started working again. His arm was so severely damaged by the swelling they had to slice the skin from his

hand to his elbow. He remained on life support for thirteen days and he had to undergo ten surgeries to try and repair the damage.
The accident did not make him lose his faith in God or the Pentecostal Signs religion he follows, Chris view is that the terrifying incident has strengthened his faith.
" When I got bitten, I told the church that this bite was serious, that it wasn't like a Copperhead bite and that I was going to need some praying and true believing in the Lord and his healing power. When I fell unconscious, they took me to the hospital, but I still give God the glory because as bad as I was, I could have died before I got to the hospital, and I didn't. It took about 40 minutes to get to the hospital once it was decided to take me, and that was about an hour after I was bitten. So all together it took at least an hour and half to get to hospital and probably another two hours before they could start any treatment."

The bite, which he received in his right forearm, was photographed and there are also subsequent pictures of him being carried out of the church by his parishioners. He contributed the photos himself and is completely open about his practice. It was precisely the fact that he had repeatedly agreed to be interviewed and provided photographic material when he had appeared in the media that made me think I would get an interview. He has happily volunteered before, even for on-camera interviews. But now the door was closed, and he was keen to know who had agreed to contribute to the content of this book.

It's a shame he didn't want to answer questions, I'm curious about his views on snakes and their ability to feel pain. He has said about the rattlesnake he stood on during the sermon that it cannot feel pain and that a human being is too light to be able to harm it by standing on it. My view is, of course, that his claim is not in any way accurate.
A snake can feel pain and can be harmed if a grown man stands on it. I think that what surprises many people is that the snake didn't bite, but it certainly didn't look like it was in good condition.
I'm also curious about his choice to seek medical attention when the bite was bad and his views on his brother and father not seeking medical attention. And of course, the book would have been better if I had another pastor to tell me about himself and his faith. But as it is, you don't always get what you want.

Photo: Steve Ludwin. Chris Wolford standing on a Timber rattlesnake.

He is now back in the pulpit handling venomous snakes, but he refuses to comment on his decision to seek medical help.

Maybe that's just the way it is? Maybe you lose yourself a little if you "act cowardly" and seek care? At the same time, it's quite fun to survive, even for a "snake pastor"? I believe that for as long as there are those who stick to the traditional way of relating to bites and care, it can feel a bit shameful to seek care.

If the snake handling phenomenon exists in the future, it will surely have completely transitioned to the fact that it goes without saying that you should seek medical attention if you have been bitten.

At least that's what I think, the future will tell if I guessed right.

"Mack" Randall Wolford

"Praise the lord and pass the rattlesnakes!"
Pastor Mack Wolford had many ways of getting the faithful crowd all fired up. He was preaching and brandishing, singing, speaking in tongues and handling venomous snakes. Wolford, who lived in McDowell County in West Virginia, belonged to *the Full Gospel Apostolic House of the Lord Jesus.*
He was a veteran snake handler, and a pastor who liked to invite journalists and photographers to show the "true face" of the faith and try to popularize it again. Mack used to coil venomous snakes round the arms and necks of his congregation and on himself during service. One of his snakes, that he called *Old Yeller*, ended up killing him with a bite on his thigh. He had already been bitten four times before.
In an interview, he talked about what it feels like to be bitten by a venomous snake.
"When it bites, at first it feels like having two red-hot nails driven into your hand, straight into your bones. And when the venom starts to spread, it feels like having your arm in an oven, it gets hot. The hand feels like it's in a meat grinder. And when you move your arm to stand up, it feels like someone is trying to dislocate your arm. It really hurts."
At the service where Mack was fatally bitten, a photographer named Lauren Pond was present because she was accompanying him to make a photo report. She shot several frames at an outdoor service as Wolford held, and then walked bare foot on, the Timber rattlesnake called *Old Yeller* he had owned for several years.
"Gone was the benevolent energy I had felt the previous year," Pond later wrote, *"and in its place an odd sense of urgency."*
As she shot some photos, *"suddenly an eerie stillness fell over the picnic site".* The cause quickly became clear for her. Wolford had been bitten. Eight hours later, he died. His family and followers respected his steadfast refusal to send for medical until it was too late. That is how strong their belief was, that his fate was solely in God's hands. Mack changed his mind at the end, but it was too late to do anything for him. In the aftermath, *ABC News* reported that the park authorities had been unaware of the nature of the event and quoted a spokesperson saying: *"If we had known about it, or if we had been asked for permission, permission would not have been granted."*

Andrew Hamblin

Andrew Hamblin was 17 years old when he first discovered the somewhat unusual Christian practice of handling venomous snakes during worship. Growing up with his grandparents at *the Free Will Baptist Church* in Tennessee, he was familiar with some of the more demonstrative evangelistic styles of Christian faith, but snakes were new to him. In his church, congregants shouted, danced and spoke in tongues - but the pastors never held up venomous snakes as they stood at the pulpit and preached.
When he visited Jamie Coots' church during a revival and saw the Pentecostal pastor handling two rattlesnakes during the service, holding them up in the air and standing face-to-face with the deadly Timber rattlesnakes, Hamblin was mesmerized. Shortly after this experience, Hamblin joined Coots' church.
It would be months before Hamblin even thought about handling venomous snakes himself, but when God called, it was time.

Within a few years, Hamblin was ordained (online), established his own congregation at *the Tabernacle Church of God* in LaFollette, Tennessee, and began handling rattlesnakes, copperheads, and cottonmouths for his congregation. Hamblin, like so many others who practice their religion the way he does, has been bitten several times. On at least one occasion, he refused medical attention after a serious bite.

When I wrote my previous book on the subject in 2022, Hamblin was still active with his snake handling, I found him on Facebook where he openly shared pictures of himself and the snakes at the pulpit. I tried to get in touch with him, but he read my messages without replying. Hamblin has always been an avid user of social media, to reach followers and gain attention and celebrity. Today, few people seek attention as many pastors did until almost ten years ago, Hamblin is an exception. This fact led me to believe that this was a man who would happily participate, but that was not the case.

Andrew Hamblin starred in the reality show *Snake Salvation* on the National Geographic Channel alongside his mentor, Jamie Coots. In 2013, when the show was filmed, Andrew was 21 years old and although Jamie was the big star, Andrew was also becoming a household name.

In November of that year, Andrew had 53 rattlesnakes and copperheads seized by *the Tennessee Wildlife Resources Agency.* At the time of the seizure, Hamblin repeatedly said he would rather be in jail than give up snake handling during church services. In Tennessee, individuals are prohibited from owning venomous snakes, but Hamblin argued that the ban violates religious freedom. His defense in court was clear and simple. The snakes were not his, they belonged to the church. Hamblin also said in his and the church's defense that state officials had no mandate to plunder a church. During the time of the investigation and trial, his parishioners ran a massive campaign to free Andrew. Signatures were gathered and social media was used extensively to build support for their cause. He was overjoyed with the jury's decision when it came, he received no penalty.
The seized snakes were taken to the Knoxville Zoo, most were in poor condition and about half died within a short time of being seized. 36 hours after having all the snakes seized, Andrew was again preaching, with rattlesnakes and copperheads. Andrew had promised his congregation that he could get new snakes, and he kept his promise. The problem with the law and the death of Jamie Coot meant that plans for a second season of the TV series did not materialize.

After Jamie's death, Andrew changed the way he held services. He no longer offered others the opportunity to hold a venomous snake during the church service. In an interview, Andrew said that he was the shepherd and responsible for what happened in the building, in response to why he had changed his services.

In 2014, Hamblin and his followers were evicted from their church because of their penchant for handling venomous snakes. Clyde Daugherty, who built the church, was at the time renting the building to Hamblin for just $267 a month. However, he eventually grew tired of all the snakes that he said were slithering around and sometimes biting parishioners or pastors.

"Too many people were getting hurt and dying, and that's a proven fact", Daugherty said.

"He was too focused on the wrong things, like attention from the media and the TV show, instead of the word of God", the property owner continued when interviewed about the eviction of Hamblin and his congregation.

In 2015, Andrew Hamblin was arrested. He was reported by his ex-wife, who had divorced him after Andrew had an adulterous affair. They had been a couple since their early teens and had five children together, although they were still young.
David Andrew Hamblin, 23, had been charged with aggravated assault and six counts of felony reckless endangerment. A judge ordered that the bond would be set at $200,000 because of the seriousness of the charges. If he was able to post bond, he was not allowed to have contact with any of the people involved, including his children.

Elizabeth Hamblin told police that as she was leaving a park with four of her children and a man called Jeremy Henegar, Andrew Hamblin came up behind them and pointed a gun at them from his vehicle. Elizabeth, Jeremy and the children left the scene and drove home, pursued by Andrew. Elizabeth Hamblin testified that when she arrived home, Andrew stopped the car in front of their house, waved a gun and fired a shot in the air. She claimed that he then pointed his gun at Jeremy Henegar and fired a second shot. The bullet missed Jeremy and went through the wall of the house into the residence. According to investigating officers, Andrew and Elizabeth's infant son were in the house when the shots were fired. Investigators found the bullet inside the home, but no one was injured. The charge resulted in his acquittal.

Andrew is not without controversy in his own circles. It is not universally recognized that he seeks the attention of the outside world. Many remember that Hamblin had previously broken a major taboo by allowing his services to be filmed. The reality show angered many more "old-school" religious snake handlers, who felt the program commercialized their religious beliefs. They spoke out against the program and have since banned all forms of filming within their churches. The critics' view is that Andrew Hamblin (and for that matter Jamie Coots when he was alive) preached for three hours and handled snakes for five minutes and the only thing that was broadcast was the handling of venomous snakes. The show simply gave a distorted view of their faith and their services. They also complained that the reality stars, in their hunger for drama, no longer waited for the "anointing of God" - an acknowledgement of the presence of the Holy Spirit - before running to their boxes with rattlesnakes.

However, Andrew is still a big name in his world. He is good at being heard and seen and likes to use social media like Facebook and YouTube. Sometimes I wonder if this is partly what the phenomenon of the book is about. Attention. In the past, you got it in church when you showed that you had a pure faith and dared to stare death in the eyes by holding a venomous snake. Today you can also reach out on TV and on social media. Everyone wants likes!

"It doesn't matter who you are, it doesn't matter where you've been, it doesn't matter what you've done, Jesus is calling!" he announced on Facebook when he opened his new church in autumn 2019. The church benches were packed from the first night.

When I was writing my book Snake church, I tried to get in touch with Andrew, but he never replied to my messages. In September 2024, I tried once again, and I sent the same message that I had successfully sent to Cody Coots a few weeks earlier. This time Andrew replied, and he also replied that he was willing to contribute to the content of the book to some extent. The reply came in the middle of the night Swedish time, and my plan was to answer him when I had a couple of cups of coffee. While I was writing a polite reply, he deleted his message and blocked me on social media. I don't know why. Either he changed his mind, or he consulted someone who thought it was a bad idea to talk to me. That's how it goes sometimes!

*

The chapter's cast of characters undoubtedly offers some colorful characters and some exciting life stories. It is certainly the case that there are pastors who are more ordinary, but of course they are not as easy to find as those who stand out. Several of those I have chosen to write about are prominent people in this world who have chosen to be seen outside their own sphere and who, through their actions, have become known far beyond Appalachia. My basic idea was to show the

reader that somewhere they are quite ordinary people, but who have an unusual way of practicing their religion. My success was probably doubtful. It is likely that many of the more ordinary people in this world have little or no interest in being heard and seen outside the church. The cast of characters was probably not as representative as I had hoped.
Some of those selected have or have had problems with the law for reasons other than preaching, but don't take that to mean that this is how snake handling pastors generally live and are. The demographic base is too small to draw that conclusion, if there had been enough pastors to write about, they would certainly have better reflected the society they live in than the current base happens to do.

I contacted some pastors when I wrote my book in Swedish on the subject. Most of them ignored me, one replied. I would probably have done the same, ignored the unknown person from the other side of the Atlantic who comes with curious questions. Of course, even if they themselves see nothing wrong with their religious practice, they are fully aware of the world view. Most are probably convinced that there is nothing good to be gained by agreeing to an interview. Many sought the limelight about ten years ago and beyond. The Coots and a few others even had their own reality series that had a fairly big impact. In retrospect, they probably realized that in the long run there was nothing good to be gained from being a celebrity, there was a hangover when they noticed that the world's view of these reality stars was not as positive as they had thought. This, combined with legislation restricting the use of snakes in the church and a world that is increasingly interested in animal welfare, has meant that today's snake handling in the church world often takes place in behind closed doors.

When I wrote this book two years after my book in Swedish on the same subject, I once again tried to contact the preachers who had ignored me. The suspicion was still there, but my "hunting skills" were a little better! Perhaps this time I was more articulate in presenting myself and my book idea?

Cody Coots responded positively this time, and it ended with the Coots family getting their own chapter. Andrew Hamblin did what he did, and Chris Wolford replied that he did not want to participate. Some have not responded at all, just like last time.

In my eyes, handling venomous snakes with your hands, drinking diluted poison, holding a flame to yourself and speaking in tongues is a bit fanatical and/or dangerous. I have a hard time understanding the purpose and benefit of it and I have a hard time understanding literalists. I guess that's the way it is with all faith, it's hard to grasp and understand for us who are non-believers.

What I can add in conclusion is that the few people I got in touch with turned out to be really nice people! Many thanks to Cody and Verlin for your help!

Phot: Steve Ludwin. A *Crotalus horridus* in its transport box. The snake belongs to Chris Wolford, pastor in West Virginia, in what is considered to be one of the last remaining churches with venomous snakes.

6. Verlin Short

The following text is from my book "Snake church" which I published in 2022. The book is in Swedish, and parts of this book are translations from it. I choose to keep the following text without changes as it describes the hunt it was to get someone to agree to be interviewed.

Verlin is quite active online. Via his group on Facebook "Snake Man of Appalachia's page", he spreads God's words, thoughts, Bible quotes and the like. On Facebook, he has almost 5,000 friends, calls himself "The Snakeman" and sometimes posts a picture of a venomous snake. Nothing can be seen of venomous snake handling, at least not if you are not friends with him on Facebook. I managed to get in touch with him even though he didn't want us to be online friends. Of those I tried to get in touch with to ask curious questions, he is the only one who responded. The others have ignored me. At the time of writing, I still do not know if there will be an interview. I have received his phone number, and the idea is that I will call to ask about twenty questions with the hope of follow-up questions. He wrote that it was uncertain whether he was going to answer or not when I called. The message I received was that I was welcome to call and that he would consult God before making the decision whether to answer or not. Hopefully I can finish the text about Verlin with his own story about why he is handling venomous snakes.
Verlin took the name "Snake Man of Appalachia" from the six-episode reality series he starred in in 2010. Snake-handling pastors were really hot in the reality show world ten years ago. The show aired on Animal Planet in 2012, but there was never a second season. In the program you get an insight into the family's life and everyday life, you get to see when Verlin catches rattlesnakes and copperheads in the wild and when he uses these during church services. The family with four

children seems to live below what you might think is a normal standard, at the same time you never know how much it is dramatized and exaggerated to make good television. When the series was filmed, Verlin was an unemployed ex-coal miner, and you probably don't get rich having a small congregation in Appalachia. Just like Cody Coots, Verlin has a wife who does not really share the belief that it is God's will that one should risk the health when in church.
The pilot episode was shot in the spring of 2010, Animal Planet ordered five more episodes, and the seven-person film crew returned to Letcher County in the fall of 2010 to film the remaining episodes. The show also addressed the issue of Verlin wanting to practice venomous snake handling legally. Over the course of the six episodes, Verlin works to obtain permits to collect more snakes and to be able to legally transport them. Verlin had previously learned that it was not a completely legal activity. In June 2008, the Kentucky Department of Fish and Wildlife charged him with 78 counts of illegally buying, selling and possessing venomous snakes. In October of the same year, a further 51 points were added. Short pleaded guilty to seven counts of illegally buying, selling and transporting reptiles. He was fined $100 for each count but avoided jail time.

*

In the beginning of October 2022, I got in touch with Verlin.
He answers me via Messenger that he will consult God in prayer, as he feels that he has been let down and deceived more than once before when he agreed to an interview.
A few weeks go by without me hearing from Verlin.
In the second half of November, I heard from him again. I do what I can to convince him that I'm after his own story, what I want to do is translate his story and have it in my Swedish book. I don't want to remove anything, change anything or add anything. I sent my questions to him so that he could write instead of having a phone interview. That way, he has a better opportunity to answer thoughtfully, answering in a live interview is not always easy and when someone is taking notes at the same time, there is always the risk of errors in the text.

I offer to put the text of my questions and his answers in English into the book and, once the book is printed, photograph the relevant pages and send him pictures.
I have gained Verlin's trust, he believes that my interest lies in the snakes and the true story of their faith practice. He's bought that I'm not out to serve up a story that's cheesier than the truth. I simply want to include a "snake pastor's" own story. However, when I write this text, it is still unclear whether he will be available for an interview. Thanksgiving 2022 I get the answer that he is still waiting for God's answer regarding a possible interview, along with his thoughts on snakes:
,,,the snakes and as of all his wild life what god made them to be and for god made and placed everything up on this earth to do what he intended for them to do as he made man in his likeness that so those who do not choose to serve him could see god is mana fest in us like wise everything he made he made it perfect and was pleased with it ,,,

I take the answer to mean that Verlin is of the opinion that snakes are not evil. They are God's animals, and they do what God intended them to do. I agree, snakes are not evil. However, I find the basis for snakes' way of life in biology and evolution, since I'm not a believer.

Another couple of days have passed, "radio silence" prevails. I'm getting a little impatient. The book is ready as soon as the possible interview is done. If the interview doesn't happen, I could just as easily send the material in for printing without it but I really want it in my book. I contacted him again, explaining that I am getting close to the deadline as I want to get the book out before Christmas. It was a short message, if it continues to drag on, I was certain that there probably wouldn't be an interview. It was evening in Sweden when I sent the message, and previously I have received answers between three and four in the morning Swedish time. The eastern United States is six hours behind us.

On December 1st, I had a message from Verlin when I woke up. He wants me to call him. He has reviewed the questions I sent him but wants to talk to me before sending his answers. He, like many others who share his way of practicing their faith, has been burned in the past. He has given interviews or appeared on television and then discovered afterwards that he, and those who share his faith, are being portrayed as fools.

In the evening, I gave Verlin a call. He immediately asked me who had been asked to participate and who was going to participate. I told him that I asked several of those who previously chose to appear in the media, but that everyone else chose to ignore the Swede with curious questions. He said that nowadays it is unusual for anyone to want to be seen or heard outside the world of the church, almost all doors are now closed. I won Verlin's trust and had a fantastically interesting conversation that went on for about 50 minutes. Verlin turned out to be a really nice man! We talked about trust; I told him that I am interested in snakes and that I also have an interest in history. I made him trust that my goal was not to hang out or blackmail people who believe in something I don't believe in, my goal is to tell a story. We compared beliefs and opinions, and both was of the opinion that you don't have to agree on a subject to get along. The outcome was that I respected his faith, and he respected my view of religion and his way of practicing his faith. We also talked about the fact that this is quite unusual in today's society, you simply don't respect those who are different or think differently than yourself. I received a standing invitation; I am welcome to visit Verlin at his home and get an insight into his practice of faith. We talked a lot about snakes, Verlin told me that he has had an interest in snakes all his life. The first time he caught an Eastern copperhead in the forest he was twelve years old.

Verlin is curious about the Swedish native snakes, and he is a little bit surprised when I told him that we only have three species of snakes in Sweden, one of which is venomous. We talked about my collection and his collection of venomous snakes. When I told him that I have White-lipped pit vipers, *Trimeresurus albolabris*, we started talking about the Vietnam War. Verlin told me about American soldiers who died from snake bites during the war, and that he calls these pit vipers "*two step snakes" **. Verlin and I don't have the same view on how toxic this pit viper species is.
*The name *two step snake* was used quite a lot by American soldiers during the war in Southeast Asia. According to the myth, you only had time to take about two steps after a bite from the green pit viper, after that life was over. I have also seen reports that kraits, *Bungarus* sp.,

was called precisely this during the war, as usual common names are confusing. I did not get the time to tell him that I had started a thinner book on *Trimeresurus albolabris** and that the relationship between soldiers and snakes during the Vietnam War was also included in the book, before the conversation moves on to other topics. Verlin seems to have an interest in herpetology; for him, snakes are not just a tool to be used in church. He told me that he takes good care of his snakes, and I have it confirmed that his snakes are well looked after between services.

*By now, the book is finished, and you can find it online. The story about the pit viper and the Vietnam war got its own chapter. The title of the book is *"White lipped pit viper – Trimeresurus albolabris".*

The conversation returns to the subject of Verlin's snakes that he has at home. He said that he had a Timber rattlesnake and quite a few Eastern copperheads at home and that in everyday life he always uses a snake hook and feeding tongs to care for and feed the snakes. He only handles venomous snakes with his hands when God calls, never otherwise. In the past, he also had non-native venomous snakes and even used these during church services. This surprised me! I wrote in the beginning of the book that I had read claims that cobras and other venomous snakes were sometimes used during worship and Verlin confirmed this. He himself had Monocle cobras *Naja kaouthia*, Puff adders, *Bitis arientans*, and pit vipers in the past, that he had used for religious purposes. Verlin said that the snakes used in worship must be venomous, but the origin of the snakes is unimportant. *"No matter where the snakes come from, they are all God's creatures."*
After we've been talking for a while, I felt it was time to ask the question about snakebites. What is Verlin's opinion about pastors who have been bitten and died? Have they had a wavering or mistaken faith because they were not saved by God? I mentioned Jamie Coots as an example. Verlin replies that he had nothing bad to say about Jamie or his faith, nor about others who have died. The answer is a slightly longer exposition that can be summarized by saying that God called on them, it was their turn to come home. He concludes that

"*God works in mysterious ways"* is a good explanation for why believers have been bitten and died.

Before we finished our conversation, Verlin told me about the one time he was bitten by a Timber rattlesnake. He was bitten in the leg, and it happened a long way out in the wild. The walk back was long, and the venom had started to work on him long before he reached his car. Verlin was closer to the hospital than home, but he chose to go home and leave his fate in God's hands. The outcome was uncertain for a while, Verlin said he hovered between life and death before he started to recover and feel better.
Then it's time to end the call and Verlin said that I'm welcome to call again. He said that he will send me the answers to my questions shortly, since we both prefer answers in writing as it minimizes the risk of misunderstanding. I assured him once again that I will not add, change or delete anything.

When I woke up the next morning, I had his reply in my inbox.

Photo next page: Photographer known.
Images from Verlin's private collection and published with his permission. Unfortunately, they are photographs of photographs which compromise the image quality. In three of the pictures, you can see Verlin handling Timber rattlesnakes during church services. In the picture in the lower right corner he is handling a *Bitis arietans*, a Puff adder. It is considered by many to be Africa's most dangerous snake, not only because it is probably the most common and widespread venomous snake, but also because of its size and strong venom. Verlin told us about the Puff adder when we spoke over the phone. According to Verlin, the snake was not his, but someone who lived nearby kept it in a terrarium. The snake's owner never free handled the snake, which to me sounds reasonable. Verlin had spoken to the owner of the snake and then felt called by God to go there and handle the snake.

Verlin says in his written reply that he was born into this practice of faith. He is 51 years old (in 2022) and his parents found the faith in 1972. The first time he handled a serpent during worship, he was only 13 years old. Today he doesn't let anyone under 18 handle a venomous snake during worship.

I asked him how someone starts with religious snake handling, and he tells me in his written answer that you start by living for God and giving God your whole heart. You should be willing to do anything God asks you to do and if God allows you to perform any of the five biblical signs, you do it. It is, according to Verlin, the power of God that allows us to do these things as a sign to non-believers. He goes on by saying that whether it's handling venomous snakes, healing the sick, casting out devils, drinking poison, or speaking in tongues, this is what Jesus commanded his disciples to do before he ascended to heaven after his resurrection.
Verlin says that he is never afraid or nervous when dealing with venomous snakes, no matter what kind of snake it is. As long as God's Holy Spirit has called on him to handle a venomous snake, there is no fear. He handles venomous snakes during worship because the Gospel of Mark says to do so. Not serving God in this way means becoming a lost soul, instead of being saved by God. For Verlin, dealing with venomous snakes is not about proving something to oneself or to other members of the congregation, it is simply about obeying God's command.

In my written questions we returned to the snake bites, what he thinks about those who are bitten and whether it can be a sign of wavering faith. He writes a longer text that begins with the bites being about God showing us humans that snakes have teeth and venom and that this is intact in the particular snake. He tells me that he himself has been bitten nine times, the first of which was almost fatal. The bite had been inflicted by a melanistic Timber rattlesnake.
This is the same bite that he previously told me about on the phone, he was bitten in the wild but went home instead of going to hospital. According to Verlin, his breathing stopped for 45 minutes and during this time he was dead. On the phone, Verlin told me that during parts of this time he could hear his father and other congregation members talking, but that they could not hear him when he tried to answer. Parishioners prayed for him, no doctors were involved, and he came back to life because God answered the prayers of the parishioners.

On another occasion, he received a bite from an Eastern copperhead that caused him incredible pain, but when he raised his hands to heaven and prayed to God, the pain went away.

Photo: Kurt Longfield. Image from Verlin's private collection, published after Verlin obtained the photographer's permission. The picture shows Verlin in the church, with a *Crotalus adamanteus*, Eastern diamondback rattlesnake.

Handling snakes or drinking poison does not make Verlin feel closer to God, the feeling that fills him when he does it is what gives him the feeling of God's presence. To stop handling snakes is not an option for Verlin as he believes in everything written in the Bible, to stop this part means that he would have to stop believing. Snakes are important to faith because in the Bible, God says that man has power over all animals and dealing with venomous snakes follows this. Handling venomous snakes has made Verlin feel closer to God, removing snakes from church services dilutes the word as you are no longer following God's signs. For Verlin, quitting venomous snakes is not an option. Although several close friends have died from handling them, he will continue to follow the Gospel of Mark. The negative effects he has experienced from venomous snake handling are exclusively about feeling persecuted by the outside world.

He has never sought medical attention after a venomous snake bite but does not rule out doing so in the future. As I have previously written, the issue of seeking medical care is something that has divided practitioners into two sides. Some believe that it is not written that one must not seek care, others believe that one should have faith in God. Verlin seems to have one foot on each side. If I interpret his answer correctly, it is up to each practitioner to decide whether or not to seek treatment.

Verlin's view of the impact of the TV series in the early 2010s is generally positive. According to him, the attention brought more churchgoers and curious people found God. There were also negative aspects, as people had to learn the hard way to be on their guard and not to trust that everyone wanted to tell the story as they thought they had told it. For Verlin's own life, his involvement in television has not changed anything.

*

Starting on the next page, you can read the questions I emailed to Verlin Short in the autumn of 2022. Exactly as they were asked and exactly the answers he gave. I have added a comma or a full stop if it was missing, otherwise it is his answer in full without any changes.

Shortly, tell me how it started for you. Were you born into the congregations that practice snake handling as a part of their practice?

"Yes, I was born into it. My mom and dad started going around serpent handling churches in 1972 and became believers and I was raised up into it."

When did you start to handle snakes?

"13 years old at a house meeting in Perry County, Kentucky, in 1984."

How do you start? Just pick up a snake or does everybody get some form of "snake education"?

"You start by living for God, and giving God your whole heart, and being willing to do all that God moves on you to do that is in his word (the Bible) and if God moves on you to carry out any of the 5 Bible signs through the anointing and spirit then you do it, it's a power from God that is not of us, but from heaven that allows us to do these things as a sign to non-believers, whether it be serpent handling, healing the sick, casting out devils, drinking poison or speaking in new tongues. (Mark 16:17-18) this is what Jesus commanded his disciples before ascending to heaven after his resurrection and is something only commanded to his followers that are the believers."

Were you nervous in the beginning and are you ever nervous now, when handling?

"As long as the Holy Ghost moves on me, I'm not scared or nervous whether it be from a juvenile copperhead or something like a bush viper from South America or cobras from India or puff adders from Africa. God has moved on me to handle a 5-foot puff adder once that had never been handled and was never took from a cage or fooled with until I went to that guys house and God moved on me to handle but with the anointing there was no fear."

Why do you handle snakes under ceremony?

"The reason why is because Jesus Christ commanded to his people in the 16th chapter of mark from 14th verse to the 19th verse. We do because he told us that he that believes shall be saved and he who

believes not will be damned (our souls would be lost if we wasn't obedient to his word)."

Is the snake handling about showing God and congregation how strong your belief is?

"The reason why I take up serpents is not to show or prove anything other then what God can do through his anointing, and I want to be obedient to his word in all that I do."

What do you think about them who got bitten? Is it a sign about a belief that is not strong enough?

"If someone gets bit, most of the time from what I saw was God was showing people that the snakes had fangs and venom and was not altered in anyway. (Ecclesiastes 10:11) this answer is for those who I've seen bitten that have not got hurt and have gotten hurt. I have been bit 9 times and I've only been hurt once when I actually died from a rattlesnake bite when I was 21 from a black rattler up on a mountain called "high knob" in Virginia. It was roughly 3 miles from the bottom of the mountain by the time I made it to the bottom I was numb from my head to my toes from the venom and I came home and trust God with the bite believing God would heal me from the bite through the prayers of my elder brothers and sisters who came to pray for me and around 8 hours in the bite I died, I felt my spirit lift from my body and the feeling was the best feeling I've ever felt in my life and my mom and dad and the others said my eyes set like a dead person and I stopped breathing for about 45 mins and the people continued to pray and 911 was never called and they prayed until God answered their prayers and eventually I felt my spirit enter back into my body the same as it left. And I felt myself take a death breathe and let the air out of my lungs and I begin to get my sight back and hearing back within 10 minutes. (Acts 28:3-9) (Luke 10:19) (Genesis 1:26). I was bitten 8 times after that and never got hurt. I was in the mountains of Kentucky digging herbs and reached down to dig some blood root and was bitten by a northern copperhead. It started burning like fire and hurting worser then one could imagine and I lifted my hands toward heaven and begin to talk to the lord about how I've served him and done all that I knowed to do through the anointing

and how I needed him to move for me and take the pain and as I layed my hands back down from praying the burning and pain had completely left me."

Does the snake handling make you feel closer to God?

"It's not the snakes or the drinking of poison that make me feel closer to God but it's the spirit that moves on me to do it that makes me feel closer."

If you stopped handling snakes, would it have a negative effect on your own belief, on your relationship with God or congregation?

"Yes, because if I stopped doing that then I can't do anything else for God because it all goes together with the 5 signs and the whole word, I have to believe and do the whole word as it is written."

How important are the snakes and why are they important?

"(Genesis 1:26) In the beginning God said he gives us power and dominion over all creatures and sign of serpent handling follows this."

Would the message to the congregation be of lesser value for them or for you without the snakes?

"Yes, because without the 5 Bible signs, 9 spiritual gifts and 5 Bible callings and the spirit moving on us to do these things it would be lesser value and not truly show the power of God and his abilities."

Under ceremony, who is allowed to pick up snakes? Everyone who wants to?

"Anybody that God moves on to do it is allowed as long as they are of age 18 through and by the spirit."

Does the fact that some people have died (or gotten very ill) from snake bites under ceremony sometimes make you think about perhaps stop handling snakes?

"Never, I have lost a lot of close friends down through the years, but nothing has made me back up on what Jesus had commanded us to do in the 16th chapter of Mark."

If you got bitten and started feeling very ill, would you seek medical help?

"As of today, at 51 years old I have not but not saying I would not but we don't know what we will do until we are in that moment."

What is your opinion about those who seek help, and about those who don´t seek help after a bite?

"My opinion is positive on both sides, because one man trusted God with the bite live or die, and the person who seeks medical help live or die does not change my opinion with my salvation because the Bible says for every man to work out his own soul salvation with fear and trembling because only I know how close I am with God and only they know how close they are with god. (Philippians 2:12)."

What would you say have been the long-time effects that the tv-shows (from about ten years ago? had on churches who practice snake handling?

"A lot of positive has come out of it and people who have visit our churches out of curiosity have witnessed the power of God and they have been negative effects by people who were not living where they needed to be with God and that's why it's so important that we are very careful with working with editors, producers and whoever may be seeking our righteousness. We must live where we need to be with God so that no negatives come from it whether it be documentaries or books or interviews."

Did the tv-shows effect your life? How?

"No, they never effected my life, I've done probably 100's of documentaries, interviews, tv shows and right books and I have never personally had a change".

What positive effects have you gotten from snake handling?

"It has made me grow closer to God and trust in his word and his spirit which moves on us. (Psalms 23)."

What negative effects have you gotten from snake handling?

"Jesus said we would be persecuted for his name's sake, I've been persecuted for how I believe but it's never had a negative effect on me that has caused me to step away from this was of believing."

Is there an obvious question that I have forgotten or something you would like to add?

"As a child growing up in holiness Pentecostal churches the power and anointing of God and the old people would always say it's better felt then told, there is no true explanation for the power and anointing of God you would have to seek God very close to be able to truly feel and understand this that we are speaking of amen. The power I'm speaking of is the power that I've seen raise the dead and cause the blind to see as we've heard Jesus do in the word, the same power that came upon Moses when God told Moses to raise his staff at the Red Sea and it parted the sea that allowed him and the children of Israel to pass through on dry land and deliver them from the hands of the pharaoh. (Exodus 14)."

Photo: Unknown. Picture from Verlin's private collection, unfortunately a photographed picture which compromises the image quality. The picture shows Verlin with a *Crotalus horridus*, Timber rattlesnake, in Arizona.

This is the end of the chapter on Verlin Short in the book I wrote two years ago.

Verlin and I have been in sporadic contact since my interviews and emailed questions in late 2022. He received a copy of my Swedish book in his mailbox as soon as it came out. Large parts of the interview with him are in both Swedish and English in the earlier book, so that he could see that what ended up in the book was consistent with what he had said and written.
He has since written to me several times saying that he thinks I should publish my book in English. In September 2024, when I started working on this book, I contacted him again. I received a short reply.

"I'll have to pray about some of the questions and get scriptures on them so everyone will know that we are not some crazy fanatic or cult I'll get back to you"

I then sent a couple more messages with some questions during the month of September but received no response. In October 2024, I was translating the chapter on Verlin into English and writing the lines you are currently reading. When it was done, I sent him a message again and asked for his email address. My idea was that for the first time he would be able to read in full what was printed two years ago. Maybe that would ease his concerns? I still had the hope that Verlin would want to contribute to the content of this book, and I thought that before I asked him to answer new questions, it would be good if he read what we have already covered.

I never got that email address; the chapter ends here.

*

When the book is printed, I will contact him again, as I promised him a copy in his mailbox. I am eternally grateful to him, without his participation my previous book on the subject would probably not have been written and therefore perhaps not this book either.

7. The Coots family

Mark 16:18: They will pick up serpents, and if they drink any deadly poison, it will not hurt them.

Photo: Friends and family to Cody Coots. Published with permission from Cody.

- *"Keep on playin'! I'm not worried at all, don't worry! God's a healer! I said Gods a healer! Amen! If I'm going to ride with the devil, he'll have to change pace!"*

These words come from Cody Coots. They are shouted out into a microphone, while the congregation loudly speaks in tongues. The small band of electric guitar, bass and drums has fallen silent despite the pastor's wishes that they should keep on playing.
His light blue shirt is covered in blood. Cody Coots continues his sermon but is rapidly getting worse. Blood is dripping from his right ear, which has just been perforated by a rattlesnake's fangs. Rattlesnakes have a venom that, among other things, causes bleeding and interferes with the blood's ability to clot; it bleeds far more than it usually does from such small wounds.

The snake is one of several that are part of his sermon, it is swung and held up in front of the congregation while Cody dances restlessly and loudly chanted out his message to the congregation through a microphone. Cody had the full-grown Timber rattlesnake around his neck when the snake's head came over his shoulder and bit him on the ear. He never saw the bite coming, his focus was on the sermon he wanted to get out to his parishioners.
In the end, he had to interrupt his sermon and was helped out of the small church by two parishioners. Behind him, several congregants were speaking in tongues or praying loudly. Cody was feeling unwell, struggled to breathe, and had started to vomit. His shirt was wet with sweat and blood. Cody asked his church members to take him to the mountaintop where God will decide whether he lives or dies. He got a lift in a parishioner's car but was taken to hospital instead.

It almost sounds like I'm trying to write a thriller, but I'm describing a real event that happened to be caught on film. Cody Coots' service on this particular day was filmed, as part of the short documentary *"My Life Inside: The Snake Church".*
As I write this, the documentary is available on YouTube, I hope it is still there when you are reading this. The fact that it was all caught on film has probably made this venomous snake bite from August 2018 one of the most famous, the rich footage gave the story wings. Snakebites are not that uncommon in these contexts, but they usually don't make the news internationally or even get much national coverage.
We can only speculate whether the parishioner went against his pastor's wishes or whether Cody actually got cold feet, since he ended up at the hospital instead of at the mountaintop. At the hospital it turns out that Cody was lucky. One of the snake fangs nearly perforated an artery in his temple, which would almost certainly have been fatal if it did.
This is where the story of Cody Coots could have ended, but he luckily survived the bite. He can continue preaching to his congregation. With the snakes, which are his and which he looks after between services.

His hospitalization is also well documented, as the filming of the documentary continued. From his hospital bed, Cody tells us that he will not stop handling snakes.
- "These snakes are part of the Bible, and that comes before everything. But if you haven't been brought up in it, you will see it as those people are crazy. I've seen a lot of people get bitten, I've seen them suffer, I've seen them come close to death. But I think if you get bitten and you have to suffer, that there was something wrong, that you didn't move when God told you to, then you went against what God said, I think."
Hospital staff are also interviewed. Radwa Martin describes the incident in brief.
- "When Cody first came to the emergency room, it was difficult for him to communicate. The snake bit close to the artery in his temple, so he's lucky he didn't bleed to death before he got to the hospital. After emergency doctors secured an airway for Cody, he was flown to

the University of Tennessee hospital where he was on life support for quite a while, they didn't know if he would make it."

A rattlesnake nearly ended Cody's life, but he survived and was able to continue his church services. Four years earlier, his father Jamie Coots was bitten by a rattlesnake during church services, and sadly he did not survive. The church was then passed on to Cody, who is now a fourth-generation pastor. Cody inherited the rattlesnake that killed his father and used it during services. The family's faith has not been shaken by the events. Cody's mum, Linda Coots, spoke out after her son got bitten in his right ear:
- *"We are a normal family; we love each other like normal families love each other. If a snake bites you but you don't get hurt, that wouldn't be a sign to those who lack faith."*

I don't know whether the documentary gives a fair picture of Cody and his faith or not. For me, who only has experience from the Swedish church, the environment is very intense, messy and loud. I am immediately unaccustomed to tongues being spoken, people spontaneously dancing and snakes being in the room. The snakes play a major role in the sermon shown in the documentary, but at the same time I have read that the sermons can last for several hours, and we do not get to see many minutes of these hours.
On the church's Facebook page, there were mixed responses to the bite. One person wished Cody well in his recovery, while another person suggested the pastor got what he deserved. Posts and comments can no longer be found on the page when I try in the fall of 2024, which I think is reasonable. I think it is unreasonable to write that someone gets what they deserve when there is a risk that the injury will be fatal. As I type these lines in, an American free handler who kept a fairly high profile online has just been bitten by an Inland taipan, *Oxyuranus microlepidotus.* I, like many others, think he was foolhardy and that a life-threatening bite would come sooner rather than later. What I have seen of this free handler online gives me the impression that he is neither a good person nor a good animal owner, but it is foreign for me to wish for someone's life to end or to enjoy someone being fatally injured. The attention seeking free handler with the taipan has nothing to do with the Pentecostal church.

The person who has gotten most attention outside the snake-handling churches is undoubtedly Jamie Coots, Linda's husband and father of Cody. Photo: Cody Coots. A photo of a framed picture, hence the quality. Jamie Coots with a Timber rattlesnake in his hands during the ceremony.

Pastorship has been passed down within the family, Jamie was the third generation and Cody is the fourth.
Jamie gained national attention back in the mid-1990s, when he and his friend, John Wayne "Punkin" Brown, appeared in the best-selling book *Salvation on Sand Mountain,* which profiled snake handlers. Brown died in 1998 from a snake bite. After his friend's death, Coots often used one of his friend's snake boxes to transport snakes in. Jamie starred in the National Geographic show *Snake Salvation*, which profiled the few snake-handling preachers in Appalachia. The series ran for 16 episodes in 2013 and made Jamie a household name. He has also appeared in many documentaries and TV shows that wanted to tell the story of his church in rural Middlesboro.

Jamie Coots probably started handling venomous snakes in his 20s. In interviews, he did not always state the same age, but the answer was always about 20 years or a few years older. According to Jamie, believers must wait to pick up venomous snakes until God says it is time. In an interview, Jamie claimed that he made a promise to God that he would never go to the hospital if he was bitten. He kept the promise for the rest of his life.
- "I believe that when it is my time to go, there is no doctor in this world who can keep me here."

His status as a snake-handling pastor meant that, in addition to holding services for his own congregation, he traveled around to other churches. While pastoring *the Full Gospel Tabernacle in the Name of Jesus*, Coots increased the number of snakes used during worship. The percentage of serious bites by parishioners during church services increased to the same extent.
Jamie Coots himself was bitten by venomous snakes eight or nine times before his fatal snakebite. The number varies in different sources and since Jamie never sought care, it is difficult to know the exact number.
A rattlesnake bite on the left arm in 1993 was, according to him, almost fatal. In 1998, he got tissue damage on the middle finger of the right hand after another bite, and lost half of that finger.

Jamie Coots was fined in 2008 for having 74 venomous snakes in his home, without permission. He then obtained the permits that the state required for private individuals to have venomous snakes at home.
Jamie was sentenced to one year of probation in 2013 for illegal possession of wildlife after traveling into Tennessee with three Timber rattlesnakes and two copperheads. He had bought the snakes in Alabama to use during the sermon and they were transported in the car in locked wooden boxes.
He had been stopped for a minor traffic offense and the police noticed the boxes. His driver's license and car keys were confiscated pending the arrival of representatives from Tennessee Wildlife Resources (TWRA). When they arrived, the snakes were seized on the spot.
Jamie pleaded not guilty and demanded the return of the seized snakes and wooden boxes. In court, the prosecutor demanded that Jamie immediately plead guilty to all charges. Otherwise, his snakes and sacramental wooden boxes would be considered contraband and immediately destroyed by the TWRA. This angered him, and when the prosecutor saw his reaction, she offered to drop all charges against the co-defendants and return the boxes if he pleaded guilty to the first charge of illegal possession of wild animals. However, the snakes would still be forfeited. Jamie accepted the prosecutor's offer and pleaded guilty.
Jamie was by this time a household name. Later that year, he published an article in the Wall Street Journal in which he demanded

that the handling of venomous snakes should be protected by the American Constitution's text on religious freedom.

*

One Saturday evening in February 2014, it was Jamie's turn to get a fatal bite. A parishioner recounts:
- *"He had one of the rattlesnakes in his hand, he came forward and he stood next to me. Everyone could see it, it just turned its head and bit him on the back of the right hand... it was done in a second."*
After the bite, Jamie dropped the snakes he was holding, but then picked them up again and continued preaching. Within minutes, however, he stopped the sermon and went to the toilet. When the ambulance arrived at the church around 8:30 p.m., they were told he had gone home, the Middlesboro Police Department said in a statement. Jamie was contacted at home but refused medical treatment. Emergency crews left his home around 9:10 p.m. When they returned about an hour later, Jamie Coots was dead.

The snake-handling pastor's son, Cody Coots, told the reporting television station that his father had been bitten eight times before, but never had such a severe reaction. The son said he had thought the bite would be just like all previous bites.
- *"We will go home, he will lie down on the sofa, he will be in pain, he will pray for a while, and he will get better. That's what happens every time, except this time it happened so fast, and it was crazy, it was really crazy."*
Cody did not expect his father to die that night. They had handled dozens of snakes together, and when Cody was a child, he had watched his father handle many more. He had seen him bitten and seen snakes sink their teeth into his flesh without producing any venom, leaving no more than a couple marks from their fangs. He had seen his father get woozy, seen him need some rest, seen his arms swell up and turn purple and seen the tip of his right finger fall off. He knew the drill. They would go back to the house and sit up the whole night and pray. God would heal him. This time the bite had a different outcoming. Jamie Coots lived to be 42 years old.

One week after his father's death, Cody took over the church. He was 21 years old, with wife and kids, a day job and now being a preacher on nights and weekends on top of it all.

*

This is where the text about the Coots family ended in the book that I wrote in Swedish in the year 2022. I had tried to contact Cody a couple of times but didn't manage to get an answer. I thought it was a shame; he was really one of those people whose story I would have liked to have included in the book. Now it didn't turn out like that, I could only write about what ended up on the news a few years earlier.

Now when I'm working on the English version in the early fall of 2024, I went online looking for him again. He doesn't call himself Cody anymore, but I managed to find him. I decided to try to contact him via Messenger, with pretty low hopes of getting a response since it didn't work out two years ago. I sat in our garden on a Saturday working for quite a while on the wording before sending the message as I thought that a bad start to the conversation may be impossible to pick up.
Many who previously appeared for interviews or appeared on television have subsequently felt cheated. Nowadays, few are interested in talking to the outside world about their religious practice involving venomous snakes. Once bitten, twice shy, I guess.

This is the message I sent to Cody, and since then I have sent it to more people I would like to get in touch with.

Hello!

My name is Rickard Ljunggren, and I am vice-chairman of the Swedish Herpetological Association. In my spare time I write non-fiction books about snakes, for my own enjoyment and because you learn a lot when you do research.
My first seven books were written in Swedish, but I wrote my latest book about the white lipped pit viper in English.

About two years ago I wrote a book (in Swedish) called "Snake church". The book is about the congregations in the southeastern United States that deal with venomous snakes during worship, but the book was just as much about the venomous snakes in Appalachia. The goal of the book was to tell about this part of the Pentecostal movement, without making fun of anyone's faith or giving the reader a sensationalized and unvarnished picture that it is fools that you are reading about. This book I have now started has the same goal, but this time I would like to get more people to participate so that the book tells the practitioners' view of snake handling.
Verlin Short contributed a lot to the content of the book two years ago, I wanted someone who told me in his own words what you do and why you do it. In order for Verlin to be sure that I did not distort what he wrote, his interview in the book was in English and he received a book so that he could read what was printed himself.
Since then, we have been in contact every now and then and he has suggested that I should write the book in English.
I have now started this work, and I would very much like to have an interview with you in the book. You can get the questions in written form so you can give thoughtful answers. What you answer and tell is what ends up in the book, without distortion.

Briefly about me:
I have been interested in snakes all my life, right now I have about 30 snakes of which about 20 are venomous. I personally choose to always use a snake hook when handling venomous snakes, but I have friends who use their hands.
I have had a Christian upbringing, but I am not a believer myself.

Best regards

Rickard

To my great joy and surprise, I got a reply fifteen minutes later!

"Hello there Good sir I would care a bit to do an interview for your book and help you out. Want to do a phone interview or zoom? Call or just send me your questions on Messenger or through text.
If we was going to do a phone call or a zoom would like to know a few days ahead of time.
So I can have my kids be really quiet. Lol I think your book would be really interesting on views of venomous snakes. Looking forward to talking to you"

This is really happening; I'm going to get an interview with Cody Coots! I got his phone number and email address and started by emailing questions to have a basis for future conversations. Three days after Cody's positive response, I sent an email with initial questions, and he responded faster than expected:

Thanks for doing this!

I think that we will start with questions in written form, and you are welcome to write slightly longer and narrative answers (my written English is better than my spoken). If there is something I miss to ask about, feel free to write and tell me even though the question is missing! If you feel that a question is too close, just write that you don't want to answer it, or answer as revealingly as feels comfortable. Some questions may seem stupid, but some readers are not familiar with the subject.
Your answers will be your answers, I will not change anything. I probably will not agree with everything (I am not a believer, and I like to use a snake hook.), but I want the book to tell the story that people from the inside want to tell. I am not a reporter who wants to write sensational headlines, I want to write a book about a slightly more unusual part of herpetology. Before printing you will get to read the text and correct errors.

Do you still preach and still use venomous snakes during worship?

"I still preach but the church I go to now does not practice serpent handing. If I'm somewhere that dose practice it and God moves on me I'll still do it. I'd love to pastor a church again that has all 5 Bible signs in the book of Mark, but the time is not yet."

*

I am surprised to read that Cody no longer handles snakes, my view is that it is such a big part of his and his family's beliefs. Further on in his answers I got to understand why, Codys view is that the outside world gives the snake handling too much attention. The snakes have some importance, but they are only a part of the service.

*

You were bitten in August 2018, and it could have ended badly. Has that incident in any way changed the way you deal with snakes?

"The bite let me know that it's Subject to happen if you don't wait is all, a noting of God."

I understand that you grew up in the part of the Pentecostal movement that places great emphasis on the Gospel of Matthew and uses venomous snakes, poison and fire during worship?

"Mark 16:15-20 KJV
And he said unto them, go ye into all the world, and preach the gospel to every creature. [16] He that believeth and is baptized shall be saved; but he that believeth not shall be damned. [17] And these signs shall follow them that believe; In my name shall they cast out devils; they shall speak with new tongues; [18] They shall take up serpents; and if they drink any deadly thing, it shall not hurt them; they shall lay hands on the sick, and they shall recover. [19] So then after the Lord had spoken unto them, he was received up into heaven and sat on the right hand of God. [20] And they went forth, and preached everywhere, the Lord working with them, and confirming the word with signs following. Amen.

So, with the serpent handing a lot of people has focused more on that than what's really important leading people to God by Witnessing and preaching.

The fire handing came from a different book in the bible

Hebrews 11:33-34 KJV

Who through faith subdued kingdoms, wrought righteousness, obtained promises, stopped the mouths of lions, [34] Quenched the violence of fire, escaped the edge of the sword, out of weakness were made strong, waxed valiant in fight, turned to flight the armies of the aliens."

*

For me as a non-believer, it is difficult to take in his answer. That literal* believers read the verses of the Bible as a truth that should shape their lives is foreign to me, for me it is only a few selected verses from a fairly thick book. It seems even stranger if the verses encourage something that, to put it mildly, can be harmful to health. At the same time, I can sometimes envy those who have faith, there is a strength and security in being a believer as long as it doesn't happen that you belong to a sect that you feel psychologically bad about being a part of. A clarification: I don't think that the faith that my book is about somehow creates mental illness in its practitioners, but of course there is a risk of physical illness.

* Within the Christian tradition, the Bible is generally considered to be the word of God. Some believe that everything written in the Bible is true, for example that the world was created in six days and that God rested on the seventh. People who believe this way are usually called literalists or fundamentalists.

The short answer from Cody about the almost fatal snake bite to his ear, that the bite was a note from God that he was in too much of a hurry to handle the snake, seems to leave a part of the question unanswered. His answer does not contain a direct answer to the question of whether the bite caused him to change his way of dealing with venomous snakes, but I interpret the answer to mean that in retrospect he does not make a big deal of the incident and that it did not lead to any change. The bite was, after all, no more than a note from God. With that being said, a note from God is probably a bigger thing than I think it is. Since Cody will get to read this text before it goes to print, perhaps he will confirm or deny my interpretation about him not changing his snake handling after the bite.

As I am sitting on my sofa writing this text, I am taking the opportunity to write a short message to Cody. It's 9 in the morning here in Sweden, which should mean that it's terribly early in the morning in the eastern United States, but he answers almost immediately.
I guess that Cody is an early riser.

Good morning!
I have started writing the chapter that you are contributing to. My book, in Swedish, that I wrote two years ago had a chapter about you and your father. That text will be translated and the beginning of the chapter. My plan is that you will get to read the text and provide feedback, before it goes to print.
I would like to have some pictures in my book. Do you have pictures of you handling snakes in church, your father doing the same, congregation members doing the same, that I can use in my book?
When the book is finished, you will get a copy in your mailbox.

Five minutes later I got the first pictures, and he wrote that he would dig for more pictures the coming weekend.

Cody with an Eastern copper-head on his snake hook. Codys real name seems to be Gregory, but I use the name Cody because that is the name he was referred to in the media.

Photo: Friends and family to Cody Coots. Published with permission from Cody.

Cody grew up in Middlesboro, Kentucky, with his sister and his parents. He told me that his upbringing was rough and that he wasn't allowed to do much as a kid. The family went to church three times a week and in my view that must have meant that there was not that much time for him to do stuff that kids usually do.

"Believing in all the bible signs set me apart. The serpents and poison definitely made me different from everyone else even though where I was raised around it. I thought it was normal"
Cody and his sister were home schooled because his sister couldn't wear pants because of the family's religious beliefs. The school told his parents that she was not allowed to wear a dress for gym class and that led to the two siblings getting their education at home. Cody notes that *"church and snakes is about all I ever knew".*
His family started practicing snake handling (people in Appalachia often say "serpent handling") in the late 70s or early 80s when Codys great grandfather started to pick up snakes during service. As a minor Cody did not handle snakes under ceremony, the congregation believed that you need to be 18 or older to handle venomous snakes in church.
I asked Cody if growing up he was trained in handling snakes, in any way expected to reduce the risk of bites and I got this answer:
"As for safety I was thought very little. Out of 9 bites, 3 was in church. The rest was in my snake room free handling and using too short of a hook or getting bit while opening the cages. Carelessness can get you hurt. In my opinion all snakes, at some point, bite even if you raised them from babies. I learned a lot about how to deal with snakes from a good friend of mine who I bought snakes from. The last time I took serpents to church I used a hook and didn't free handle them. I just held them by the tail, so they didn't slide of the hook. At times I didn't even touch their tails."
That he used a snake hook on the last occasion was a little surprising. An obvious follow-up question for me is whether the snake still fulfills its purpose if you don't hold it with your hands. And if it is the case that you might as well use a snake hook, why risk your health?
I got a quick response from Cody. He answers that the last time he handled snakes in church was in September 2023 and the hook was just used when packing the snake afterwards. He tells me that he, his wife and about ten others handled the copperhead with their hands, and I got a few pictures that were taken that day. In our conversation about the use of a snake hook instead of free handling we came to discuss the famous bite that hit the news 2018. It turns out that Cody got bitten in 2015, but the bite did not hit the news until the documentary was released three years later!

Photo: Friends and family to Cody Coots. Published with permission from Cody.

In the picture you can see Cody dancing with a copperhead. Below you can see Codys wife Cassy with the same snake. Both pictures were taken in September 2023.

In September 2024, Cody and I are in contact quite often via Messenger. You can tell he's a nice guy, who's very helpful to others. In addition to taking the time to answer all the questions I bombard him with, he advises me on others I should contact.
As I have already written several times in the book, I am not a believer. However, I am genuinely convinced that almost all priests and pastors are warm-hearted people who want to help others and who want to do good to those around them.

I asked Cody if he wanted his children to follow in his footsteps, and the answer to this question confirms what he said earlier, that snakes play a smaller role than the outside world thinks.
"I hope my children do when their 18 but I want to teach them that there are more important things when it comes to serving good than serpent handing, even though it is right and if used and God's time can be a sign to the unbelievers."

When asked if his children are trained in snake handling, I got an answer that could just as easily have been my own.
"Training for now is never touch any snakes unless I'm around to verify it's not venomous. When they are 18, I'll show them how to use a hook always and all the snake safety I know."

For his part, Cody started helping with the snakes when he was 16 years old. His father took the snakes out of their cages and put them in a barrel, so that Cody could clean the cages and change the water in the water bowls. When he was 18, he started doing the work on his own and then he also started feeding the snakes.
I asked Cody where the snakes are kept between services. Right now, he doesn't have any snakes, but he does have a separate building where the terrariums always are. One rule that was in place growing up and is still in place in his house is that there should be no venomous snakes in the home. The room where he has (had) venomous snakes is shown in the documentary released in 2018.

The service and the snake handling

I have repeatedly encountered the claim that the outside world gives snakes more space and importance than they actually have in these churches. I asked Cody to briefly describe what a ceremony usually looks like.

*"The service starts just like any other service you've ever seen. The music is very loud!"**
Usually, we start with prayer requests. We sit down and pray, then we sing some songs. After that we have an alter call for the lost and after that the service is over and it's time to go home. The snakes may or may not be handled in a service, sometimes months go by before they are handled. Really, they just stay in the box until someone is moved on by God to get them out. If someone is moved on, they will remove the lock on the box and reach for the snake and get it out. Anyone who is moved on by God can handle them, it's not something that you get training for beforehand.

To me, the serpents don't bring much, they are just a small part of Mark 16. You can preach without serpents; you preach to uplift and encourage people and to get people saved.

Snakes interest people outside the church more than the church itself. But if that is why they are here, it is still an opportunity to tell them about Jesus Christ and how he died for all our sins. To me, the snakes are just a small part of the church. I like to keep them and care for them. I believe they are beautiful creatures."

* The high sound is something I'm not used to from the Swedish churches, but I don't know what the volume is like in Swedish Pentecostal churches. From what I've seen on TV, American churches generally are louder. If gospel is being sung, it's really loud and the preacher uses the full volume of his voice to deliver their message. The message and the songs are delivered with great power!

Is anyone allowed to handle snakes during worship? Do you deny someone because of young or old age? Is someone denied it for health reasons (that you might not survive a bite if you have more fragile health)?

"There are two reasons to deny someone handling serpents. They are not over 18 or God speaks to you and says not to hand them the snake."

I ask the obvious question, whether it is important that the snakes being handled are venomous. As I guessed, the danger serves a purpose, without the venom there is no contact with God.

"If the serpents weren't venomous, it would not be much of a sign. With the venomous snakes, I've seen God shut their mouth even though they tried to bite, and I've seen people get bite and it did not hurt them."

How do you feel inside when you, during a church service, handle a snake that could kill you?

"I can't explain the feeling but in church when the Holy Ghost moves on you to handle a serpent it's like you're not even in your own body. Sometimes I've came back to myself and didn't remember what happened."

Do fire, venomous snakes and poison have the same purpose and function during worship, so it doesn't matter which one you choose to use? If so, how do you know which to choose? Or is it God who chooses for you?

"All the bible signs that we do serve their purpose we can't choose which one to do but whichever the Holy Ghost tells you to do. In any service we are there to let God use us as he sees fit, we are there to be a help and get help all through Christ. If you get to thinking it's all about you and you try to choose what sign you're going to do, you'll end up hurt or killed."

Let's move away from the services and talk about television instead. We start with the National Geographic show *Snake Salvation* which aired in 2013. Cody's father Jamie starred alongside Andrew Hamblin in the 16-episode season. The show had a huge impact.

Storyline:
In the hills of Appalachia, Pentecostal pastors Jamie Coots and Andrew Hamblin struggle to keep an over-100-year-old tradition alive: the practice of handling deadly snakes in church. Jamie and Andrew believe in a bible passage that suggests a poisonous (!) *snakebite will not harm them as long as they are anointed by God's power. If they don't practice the ritual of snake handling, they believe they are destined for hell. The pastors must frequently battle the law, a disapproving society, and even at times their own families to keep their way of life alive.*

Exciting television, in other words, and the focus is entirely on the handling of poisonous snakes. I'm a bit curious about what it was like with a camera crew following you in everyday life and in church. Cody is asked a bunch of questions, some of which are quite intimate, which he kindly answers.

About ten years ago, several pastors participated in television series where you could follow their everyday life and their services. How do you see this today?

"As for the TV show I believe they were only focusing on snakes, they could not care less about teaching people about Jesus. They would film a three-hour service with singing, shouting, dancing, praying for people. By filming about an hour of preaching and five minutes of serpent handling, they put together an episode that was 90% about the snakes."

Your father Jamie Coots became famous around the world when he appeared in the TV series Snake Salvation in 2013, what was it like as a 21-year-old to have cameras following you and your family?

"Honestly it sucked having cameras around all the time, filming all day and then doing interviews afterwards. It was 16-to-17-hour long days!"

Was it a good forum for you (snake handling preachers) to appear on?

"At times I regret ever doing this, but I guess in some ways It helped get people to kind of study the bible and get to know Jesus Christ."

Did the TV series change your life in any way?

"It did change me some. I hate seeing or hearing myself at all, but it has made me not want to be a jerk like I was back then"

Do you think the TV series gave a fair picture of your family and your faith?

"No! I feel they just made us look like everyone already thought us to be. A bunch of dumb mountain folks and crazy snake handlers as they called us!"

Did the series give an accurate picture of your religious practice, or did it give viewers a distorted picture?

"It did not! Like I said earlier it may be months before the serpents were handled and that's the only part they showed of the service. Even if it took two months for it to happen that's all they showed!"

Did you feel cheated in any way when you watched the series?

"I don't believe I was cheated in any way; I just wish I would have stayed away from the cameras in general."

Did the tv-shows create problems for you afterwards? (unwanted attention from government or animal rights associations)

"It opened doors for the fish and wildlife to keep a close eye on everything the snake handlers did."*

* The United States Fish and Wildlife Service (USFWS or FWS) is a U.S. federal government agency within the U.S. Department of the Interior which oversees the management of fish, wildlife, and natural habitats in the United States.

We'll leave *Snake Salvation* and move on to talk about the documentary that starred Cody. As I've written before, you can still find it on YouTube at the time of writing. Two episodes of about 15 minutes each. Again, the focus is on the snakes and especially the bite that Cody got during the filming. In the world's defense, it should perhaps be said that it is actually exciting television with dangerous snakes. A documentary about Pentecostals who don't handle venomous snakes, drink poison or hold an open flame to their bodies wouldn't get anywhere near as many viewers, and viewers mean advertising revenue.

You yourself had the main role in a documentary in 2018 where you were unfortunately bitten. The documentary is on YouTube and clips of the bite and your subsequent health have spread around the world. What is your view on the documentary, does it give a fair picture of you and your faith?

"At the time of that recording I was very arrogant and cocky! I was doing it for all the wrong reasons! I wanted to make a name for myself as a snake handler. It did just that, it proved I was ignorant and should have been waiting on the movement of God that day."

You chose to seek treatment after the bite that was in the film, which I see as part of the change that has taken place, that more people are seeking treatment as it is not written in the Bible that you must not do this. At the same time, some older people think that it is wrong to seek care, that you are not putting your fate in God's hands, as you are supposed to do. What are your thoughts on this? Where do you stand on the question of whether to seek treatment or not?

"As for treatment I see nothing wrong with it nor everyone's faith is the same. I will say though in my case the doctors knew who I was and how I was in church. When I got bitten I was treated really bad before everything went black."

Would you have done anything differently today?

"Today I see snake handling very differently. I don't believe the snakes are supposed to be packed every time you go to church. When I had snakes a few years ago I fed and took care of them and it might have

went up to six months before the Lord told me to carry any to church. So, what would I do differently today? Today I would have been led by the Spirit before handling snakes, and have nothing to do with cameras "

What is your view today that your father did not choose to seek treatment when he was bitten in the thumb by a Crotalus horridus in 2014?

"My dad didn't go to the doctor because he made a vow to God that if he ever went to the doctor for a snake bite he would never go back to church. And he took that vow very serious "

I read somewhere that you inherited your father's snakes, including the snake that gave him a fatal bite. Is this true and if so, what were your thoughts when handling this particular rattlesnake?

"The feeling When God moves on you to handle, that is a peaceful feeling. You know that you're going to be okay. You do not have to worry about watching the snake's head because you know when you're moved on. And you will be okay. "

Besides the bite that was featured in the 2018 documentary, have you been bitten more times?

"My last bite was from a western diamondback on the hand being carelessness and not using a hook. I could feel my legs or feet I was screaming in pain ever time my friends would try stand me up I'd black out."*

* The Western diamondback rattlesnake, *Crotalus atrox,* is probably responsible for more deaths than any other venomous snake in the USA. Its venom is not as strong as that of some other rattlesnakes, but it is a large snake with large venom glands.

I have come to understand that today people avoid the media and that it has largely become a closed community. Is that so or have I got the wrong picture? If it is the case that much nowadays takes place behind closed doors, why has this change come about?

"People stopped dealing with media because so many people were out to make a dollar. Even if that meant trashing the people who handled

snakes! There were very few people in the snake handling world that tried with the media, and it always ended up being negative. Even to this day it seems that way."

*

The interview ends with us talking about the snakes and their well-being. What is Cody's view of how the snakes are affected by the handling and crowding of the churches? I am personally convinced that the snakes would have been better off not being handled. In my opinion, it is unnecessary stress for snakes to be close to humans, whom they see as a potential threat. The more impressions the snakes get, especially people's scent and movement, the higher the stress level of the snake. At the same time, I am not always better myself. I sometimes give lectures on snakes at libraries in Sweden, and I always have snakes with me that I show and some snakes that the curious public can touch or hold. I do this a few times a year, although I am absolutely convinced that the best thing for the snakes would be to stay at home. My view is that if I reduce people's fear of snakes, our snakes in the wild will do better if they happen to encounter a human.

I have seen a clip where a pastor was stepping on and standing on a Timber rattlesnake while preaching, and the reporter asked if that did not harm the snake. The answer given was that it does not harm a snake to be run over by a car and that it is impossible to hurt them with your body weight. What is your view on that statement?

"No comments on that video brother."

How do you feel about the claim that the snakes used during worship are often in such poor condition (malnourished and/or dehydrated) that they are almost unable to bite?

"On that question I've seen a lot of people who didn't take care of their snakes unfortunately. I do not agree with that! These snakes, although they sometimes are a part of the service, needs to be treated with respect and care like you would any animal you own. Most times when people have been bitten, the snake was handled by hundreds and the snake was weak and hanging head to tail. I personally like my

snakes healthy fat and full or spunk. One thing I know personally is that if you don't wait on Jesus to tell you to do it you're going to get bit as soon as you open that lid and stick your hand in that box."

How do you react to animal rights organizations' claims that the snakes are subjected to unnecessary stress and animal cruelty when they are included during religious services?

"I've seen people be ruff with snakes in services, but you don't have to be. They can be handled with care and not abused."

Do you get into trouble with the authorities because you use snakes during worship?

"You can get in trouble, but I never personally have."

Are the snakes you use wild-caught, or captive-bred?

"I've had both wild caught and captive-bred that I took to church."

Do you only use American venomous snakes, or do you also use exotic venomous snakes?

"Due time fish and wildlife only local snakes are used nowadays."

Do you feel persecuted in your religious practice or is it a matter of a curious outside world always having a skewed perspective (too much focus on the snakes)?

"Nowadays I don't feel persecuted when people do get worked up over it. When I explain that the snake handling is just a small part, and that I see it differently than I used too, they seem to have a better understanding."

What is your view of how the snakes are affected by the handling in church, many people moving close around them and different smells from people and sometimes smells of poison or fire.

"I believe the snakes are not affected depending on how they are treated. And as for people being close together if you're not moved on by the Holy Ghost to handle serpents you could be putting everyone around you at risk of getting bit. I've seen them strike out at people from another person's hands. The poison doesn't really have a smell

but thinking about it they probably don't like the smell of the smoke from the fire."

Before I considered this chapter ready for proofreading, I let Cody read the text I wrote. I wanted him to be comfortable with the result and not feel that I biased his answers. Cody has approved the text and the small changes I have made to his answers have been purely for the sake of clarification for the reader.

Photos: Friends and family to Cody Coots. Published with permission from Cody.

Upper picture, Cody with a Timber rattlesnake in his hands.

Lower picture, Cody with various venomous snakes in both hands.

8. Timber rattlesnake

Family	*Viperidae*	Vipers
Genus	*Crotalus*	Rattlesnakes
Species	*Crotalus horridus*	Timber rattlesnake

Photo: Dr. Edward J. Wozniak. Timber rattlesnake, *Crotalus horridus*. Picture taken in the south of Georgia.

The genus name, *Crotalus*, is derived from the Greek word "*crotalon*" meaning "rattle", and the species name, *horridus*, meaning "horrid" or "terrible", was bestowed (some would say in error) by Linnaeus in 1758, who examined only a preserved specimen.

For many people it is probably an unpleasant surprise to suddenly hear the sound of a rattle! And there is something special about this warning sound! Rattlesnakes are snakes that probably everyone knows, even those who have absolutely no interest in nature or snakes can usually name them and cobras.

The Timber rattlesnake may be considered to play the leading role among the snakes in this book. Although Eastern copperheads and Northern cottonmouths (called Water moccasins in Sweden) are also used in this part of the Pentecostal Church, it seems that it is the Timber rattlesnake that is most frequently used. Whether this is due to the impressive size of the species, church tradition, local availability (most likely) or the fact that they may bite less frequently, I do not know. The answer I have received, from the few practitioners who have chosen to answer questions, is that they use the snakes they come across. The important thing is that it is a venomous snake, not what species it is. Phil Dunning, who works as a biologist/herpetologist, gave me this answer:
Copperheads would be the most common of the venomous snakes in the regions these churches are practicing, but due to their secretive nature Timber rattlesnakes at den sites or at gestation sites may just be easier to capture in numbers.

Rattlesnakes are pit vipers (subfamily *Crotalinae*) found only in the Americas. Most of them belong to the genus *Crotalus*, but there is also a much smaller genus of North American rattlesnakes, *Sistrurus*, (often referred to as the dwarf rattlesnakes) that contains the smaller Pygmy and Massasauga rattlesnakes. Rattlesnakes are sometimes carelessly referred to as a single species, but the genus *Crotalus* actually contains 55 species while *Sistrurus* has 3. As far as I know, no rattlesnakes of the genus *Sistrurus*, being small and typically uncommon in areas where these churches mostly occur, have ever been used by Pentecostals for their snake handling services.

Some species of the genus *Crotalus* can be quite large and bulky, but there are also many rattlesnakes that never grow close to a meter in length. The main characteristic of the rattlesnakes is of course the namesake appendage at the end of the tail, which is made up of hollow segments of keratin, loosely attached in a string so that they beat against each other when the tail is vibrated to produce a droning or hissing sound that can be quite audible at a distance. Like most vipers, rattlesnakes give birth to live young. Neonate rattlesnakes, often referred to as pups, are born with a single keratinous segment on the tip of the tail known as the "button" which is incapable of producing any sound. From there on, each time the snake sheds its skin, a new segment is added to the base of the rattle. The original button remains at the tip of the rattle unless it is broken off, which is not an uncommon occurrence. By the time rattlesnakes are adults we almost never see the original button. Also, it is not uncommon to find a rattlesnake who has no rattle at all.

When a rattlesnake feels threatened, it rapidly vibrates its tail, and the characteristic rattling sound is heard. As with all venomous snakes, the primary use of the venom is to immobilize and subdue its prey. In the case of many species, especially among vipers, the venom also assists in initiating the digestive process. Being primarily made up of protein and various lipids, venom is biologically costly for the snakes to produce, and despite the doubts and misgivings of most people, they really have no desire to waste their venom by biting us and prefer to avoid altercations altogether. A rattlesnake has nothing to gain and everything to lose from engaging with any animal that is not prey, and left to themselves, they will go to great lengths to avoid doing so.

The rattlesnake rattles when it feels threatened in the hope of warning or scaring away an unwary intruder and avoiding an interaction that is likely to be of no benefit to either party.

Many people believe that the age of a rattlesnake can be determined by the length of its rattle, but while a trained biologist might be able to roughly estimate the age of a snake based on the size and appearance of the rattle, there is no direct correlation between the number of segments in the rattle and the snake's age in years. As mentioned previously, a new segment is added to the rattle every time the snake sheds its skin, and a young, rapidly growing rattlesnake may shed as many as three to four times just in its first year, such that a rattlesnake that is only one year in age could have three or four segments in its rattle. Again, as stated previously, rattles are often broken off.

Rattlesnake venom is largely haemotoxic, causing haemolysis (red blood cells destruction) and tissue damage (necrosis). The Mojave Rattlesnake, *Crotalus scutulatus*, and some of the canebrakes (lowland Timber rattlesnakes) are notable exceptions that have pronounced neurotoxic components to their venom.)

The strength of rattlesnake venom varies from species to species. The western Mojave Rattlesnake, *Crotalus scutulatus*, and some of the canebrakes (lowland Timber rattlesnakes) are notable exceptions that have pronounced neurotoxic components to their venom. The potency and constituency of the venom varies from species to species and in some cases within the same species or at different life stages. Some species have venom that isn't stronger than that of some less potent European viper species, while other species have a bite that is life-threatening. A bite is very painful, and deaths do occur, but bites on humans from Timber rattlesnakes are quite rare. The Timber rattlesnake is one of the most dangerous snakes in North America, due to its long fangs, impressive size and the large amount of venom it is able to deliver. This is somewhat balanced by its relatively mild manner. Before they bite in defense, they usually have been warning for some time. The deaths that occur after bites during worship are probably most often caused by Timber rattlesnakes. Those who seek care and get it in time usually survive, but many choose not to seek care. USA has good healthcare, but a bite can be expensive as the country does not have publicly funded healthcare. After being bitten a first time our bodies quite often build up an allergy to the venom.

Timber rattlesnakes occur in two very distinct geographic populations. The upland, or mountain, Timber Rattlesnake ranges up the Appalachian mountains from Georgia to New England, and farther west to southeast Minnesota and down the Mississippi Valley (though they have been extirpated from many areas). The lowland, or coastal version, often referred to as the Canebrake rattlesnake is found in the southeastern Coastal Plain from extreme southeast Virginia down to northern Florida and across the Gulf states to south-central Texas. Until fairly recently the Canebrake Rattlesnake was treated as a subspecies of the Timber Rattlesnake, *Crotalus horridus atricaudatus,* but that taxonomic distinction is no longer formally recognized. The name Canebrake still lives on, and the Timber rattlesnake will perhaps always be called that in some parts of its range.

The Timber rattlesnake is a quite large species, both for being venomous snakes and for being rattlesnakes. In terms of length, they are usually said to be up to 150cm long, but usually they are around 110cm. The reported record length is 189.2cm (Klauber, 1956).

Large specimens can reportedly weigh as much as 4.5kg, which is a lot for a venomous snake. Usually, it is snakes that constrict their prey that grow a little larger and heavier.

Timber rattlesnakes have a pattern of black or sometimes dark brown crossbands (chevrons) shaped somewhat like jagged depictions or birds or bats on a variable on a background that can be yellowish, beige or greyish, and sometimes pink in the southern Canebrake populations. The crossbands have irregular zigzag edges and may be V- or M-shaped. Sometimes there is a rust-colored stripe along the vertebrae, see picture on next page. On the belly they are pale yellowish with some smaller, dark spots. Timber rattlesnakes are quite grey when they are born, both pattern and ground color are greyish.

Photo: Wikimedia Commons / Public Domain. Image from Mark Catesby's *The Natural History of Carolina, Florida and the Bahama Islands*. As you can see above the drawn snake, it is a picture of a *Vipera caudisona* as the species was called when the picture was drawn in 1743. *Vipera caudisona* is now a junior synonym of *Crotalus horridus.*

Photo: Dr. Edward J. Wozniak. *Crotalus horridus.*
Adult Timber rattlesnake with the stripe at the top of the back that the species has in some parts of its range. The name Canebrake is still commonly used in both the species region and in the reptile hobby. The name is derived from its natural habitat and refers to areas densely populated with Canebrake or Giant reed. These habitats often provide an ideal environment for this species, which is typically found in forests, swamps, and areas with thick underbrush. The name reflects both the snake's ecological preferences and its association with regions where tall grasses and shrubs are prevalent.

Photo: Phil Dunning. Newborn *horridus* about to shed for the first time (known as the post-natal shed, or PNS).

Like many other pit vipers, Timber rattlesnakes and cottonmouths (and to a lesser degree, copperheads) exhibit parental care of their neonates, watching over them for about the first ten days, until they have their post-natal shed.

In the mountain populations of *Crotalus horridus*, farther north, there are two basic color phases - light or yellow, and dark or black. Some specimens are very dark and almost completely black. Really dark pigmented snakes are also common in the northern range of *Vipera berus* where I live. The difference is that the vipers in Sweden often become completely black but the American rattlesnakes that this chapter is about do so more seldom. Some older specimens of black-phase *Crotalus horridus,* especially in the northernmost parts of their range, can be entirely black, just like adders.

In everyday speech, these snakes are said to be melanistic, both when talking about *Vipera berus* (European adder) and *Crotalus horridus.* Melanism is the congenital excess of melanin in an organism resulting in dark pigment. It gives the animal a darker, or all-black, pigmentation.
No specimens of either the Timber rattlesnake or the European adder are born really dark or completely black, and thus they are not genetically melanistic - it is just a regional color variation, an ontogenetic darkening with age. They are born with the same colors and patterns as others of the same species (as in the photo above) and gradually darken with age. Truly melanistic animals, such as the black panther, are born entirely black even though their parents might be heterozygous and exhibit normal coloration (normal phenotype).

Photo: Rickard Ljunggren. Melanistic Adder, *Vipera berus*. Picture taken two kilometers from my home, in the south of Sweden. I use the term melanistic, because that's what "everyone" calls our black vipers. If you ask me, an all-black adder is the most beautiful thing you can find in Swedish nature and one of the most beautiful snakes in the world.

Photo: Phil Dunning. *Crotalus horridus* at overwintering site.

Hibernation site (often called "den") with several Timber rattlesnakes. As in Sweden, the snakes in the region need to hibernate at frost-free depths underground. The season for hibernation varies with the local climate. Rattlesnakes in the southern parts of the species' range hibernate for a shorter time, grow faster, reproduce at an earlier age and have shorter reproductive intervals than rattlesnakes in the northern parts of the range [11]. Mating occurs from mid-July to the end of October, with males locating females for mating using their sense of smell and previous experience.

Timber rattlesnakes have a fairly long reproductive interval and rather late sexual maturity. Sexual maturity is related to body size and condition and varies within the range. They reach sexual maturity between 4 and 11 years of age, with males reaching sexual maturity faster than females. In northeastern New York (state), males reached sexual maturity at an average age of 5.3 years and females at 8.3 years. In Appalachia, males reached sexual maturity at 4.4 years and females at 6.8 years. Females produce litters at 2-year, 3-year or 4-year intervals. In Appalachia, the average reproductive interval is 3 years, which means that a female will have three to five litters in her lifetime. The lifespan of the wild Timber rattlesnakes is usually 10 to 20 years but can be up to and even above 30 years in extreme cases.

Ovulation and fertilization take place in spring, four to six weeks after they emerge from hibernation. The snakes are gravid during the warmest time of the year, from late July to early October. The timing of the birth can vary greatly from year to year within a region depending on the weather. Timber rattlesnakes, like copperheads and cottonmouths, are ovoviviparous. This is a bridging form of reproduction between egg-laying and live-bearing reproduction. This means that these animals have embryos that develop inside eggs that remain in the mother's body and hatch there. Litter size varies from 5 to 22 and averages 6 to 8. Females do not care for the newborns but have been observed to remain within one meter of each other for up to two weeks after giving birth. As with other snake species, females can store sperm for long periods of time.

Juvenile Timber rattlesnakes have a lower survival rate than adults. Causes of death for newborns include predation, lack of suitable prey and lack of suitable burrows for shelter. Not all neonates feed during the period between birth and first hibernation. This is not usually a cause of death, but starvation after first hibernation can be. The most common cause of death of adult specimens is encounters with humans and the impact we have on nature. This can include hunting, deliberate collection of gravid females to reduce populations, and destruction of habitat and hibernation sites. An undisturbed hibernation site is used year after year, the snakes return to the same place every winter.

A hibernation site left alone may have been used for many hundreds of years, in some cases for several thousand years [12]. The same applies to hibernation sites for other snake species.
Non-human causes of death include starvation, predation and freezing to death during hibernation. Males have an increased risk of getting in the way of predators during the mating season, as they move around in search of a mating female.

A fairly large part of their life is spent overwintering, and the length of hibernation varies with the length of the winter.
In the northernmost part of the Timber rattlesnake's range, overwintering is about 7.5 months and in the southern parts about 5 months. In the Appalachian region, wintering averages just under 7 months, from October to the end of April or beginning of May. The start and duration of wintering varies from year to year, depending on the weather. When the frost comes and it is too cold to hunt, eat and digest food, it is time to move to the hibernation site.
Timber rattlesnakes overwinter together, even with other species of snake. Wintering sites have been found with up to 100-200 Timber rattlesnakes with a record of 250 from 1906 [13]!
In addition to the arrival of spring, the end of the hibernation period seems to be determined by the snakes' innate biological rhythm.

Timber rattlesnakes move seasonally between winter and summer locations, with each location meeting the seasonal needs of the snake. These two locations can be quite far apart but also be essentially the same area. Males move over a larger area than females, gravid females stay close to the wintering site. The wintering site offers safe resting in frost-free depths, the summer site offers food and a suitable place to thermoregulate and mate. In autumn, they return to the same wintering site as the previous year. Snakes orientate themselves by following scent trails from the skin secretions of other individuals. The journey back and forth between summer and winter sites is always the same, as long as nothing in nature is changed or destroyed.

Photos: Phil Dunning. Timber rattlesnakes at a gestation site in Clinton County, Pennsylvania.

Timber rattlesnakes are found in the eastern half of the United States. The species occurs in an area from southern Minnesota southwards to eastern Texas. The range extends eastwards to the southern parts of New Hampshire and southwards to northern Florida. The historical range of the Timber rattlesnake includes southern Ontario and southern Quebec in Canada but is unfortunately considered extinct in Canada since 2001. In general, this species is found in hardwood forests in rugged terrain. During the summer, gravid females seem to prefer open, rocky ledges where temperatures are higher, while males and non-gravid females tend to spend more time in cooler, denser woodland with a more closed "forest roof". During spring and autumn, Timber rattlesnakes are diurnal, but during the summer when temperatures are at their hottest, they switch to being dusk and night active to avoid the worst heat.

In the Appalachia's, males actively forage from early spring until the mating season begins in summer. During the mating season, males rarely feed but resume foraging in early autumn. Non-gravid females feed continuously from mid-May to the end of September; gravid females rarely or never feed during the latter part of their term. All snake species in Appalachia stop feeding when autumn arrives. When hibernation starts, the stomach must be empty, a cold snake does not have a functioning digestive system.

What's on the Timber rattlesnake's menu differs depending on gender, season and body size. Young Timber rattlesnakes like to eat various species of mice while larger specimens prefer rats, rabbits, and various squirrels, which are more palatable to a larger snake. Their hunting strategy is to lie motionless and wait for a prey animal that has not detected the snake to come too close. The snake is aided by its eyesight, a tongue that detects scents, the ability to sense vibrations in the ground and even the ability to "see" infrared radiation*, as its prey has a body heat that is higher than the ambient temperature. Therefore, they emit a more intense thermal radiation than their surroundings and the rattlesnakes use this thermal radiation to "see" their prey. This combination of tools means that the snakes can also hunt in the middle of the night, when it is completely black.

* The pit in pit viper's is what allows them to see the infrared radiation.

The prey is killed by a bite where the snake injects its venom. The prey is released immediately after the bite but doesn't get far before the venom takes effect. As well as envenoming the prey, the venom breaks down the prey's body tissue, making it easier to digest. Hunting usually takes place after nine in the evening and lasts until morning. A Timber rattlesnake that hunts spends up to nearly ten hours a day in an ambush position.

All animals have their place in the food chain. The Timber rattlesnake is eaten by birds of prey such as hawks and owls, but also by other snakes, foxes, coyotes, bears, weasels and skunks. Nature always keeps the balance as long as we humans do not interfere. The predators will never be too many, the prey will never be too few.

Human habitat destruction and hunting have the biggest negative impact on Timber rattlesnake populations. Unfortunately, the same can probably be said for many animal species on our planet.
Many Timber rattlesnake populations have declined or become extinct since the European colonization of North America. Surveys conducted from the 1940s to the 1990s have shown widespread declines, mainly due to human activities. These activities include destruction of the snakes' habitat when we humans' clear forests to expand our own living space, but also the outright hunting of rattlesnakes and the collection of rattlesnakes for the purpose of localized extermination. There have even been cases where hibernation sites have been blown up, poisoned with fuel or filled in. My belief is that the best thing we humans can do in nature is to do as little as possible. If we remove one part of the food chain, it will have an effect elsewhere in the chain. All species in nature fulfil a function, we should not try to remove or add anything to that chain.

9. Eastern copperhead

Family	*Viperidae*	Vipers
Genus	*Agkistrodon*	American moccasins
Species	*Agkistrodon contortrix*	Eastern copperhead

Photo: Dr. Edward J. Wozniak. Picture taken in Harris County, Texas.

The genus name, *Agkistrodon*, comes from the Greek words "*ankistron*" (fishhook) and "*odon*" (tooth) and is a reference to the snake's fangs. The species' name, *contortrix*, comes from the Latin word *contortus* meaning twisted, intricate, complex and refers to the snake's pattern. In my opinion, this is the most beautiful snake species in North America!

In English, this particular species is called Eastern copperhead and there is a closely related species called Broad-banded copperhead, *Agkistrodon laticinctus*. Historically, there was one species of Copperhead, with five subspecies. The change in taxonomy has not been universally adopted and the old names of the five subspecies are still more common than the current division.

For the five subspecies of Copperhead, three subspecies were found to be monotypic (taxonomic term meaning that a group has only one biological type) and all three species belong to the species of this chapter. The other two were grouped together as *laticinctus*, the Broad-banded copperhead.

Former taxonomy:	Current taxonomy:
Southern copperhead *Agkistrodon contortrix contortrix*	Eastern copperhead *Agkistrodon contortrix*
Northern copperhead *Agkistrodon contortrix mokasen*	Eastern copperhead *Agkistrodon contortrix*
Osage copperhead *Agkistrodon contortrix phaeogaster*	Eastern copperhead *Agkistrodon contortrix*
Broad-banded copperhead *Agkistrodon contortrix laticinctus*	Broad-banded copperhead *Agkistrodon laticinctus*
Trans-Pecos copperhead *Agkistrodon contortrix pictigaster*	Broad-banded copperhead *Agkistrodon laticinctus*

American moccasins, *Agkistrodon*, are a genus of vipers found only in North America, from the northeastern and central United States down to northern Costa Rica. Currently, eight species are recognized, including this chapter's Eastern copperhead and the next chapters Northern cottonmouth. All eight species are venomous. Copperheads and cottonmouths account for many of the venomous snake bites in the wild in the United States, but they have a milder venom than many rattlesnake species and deaths after a bite from *Agkistrodon* species are quite rare.

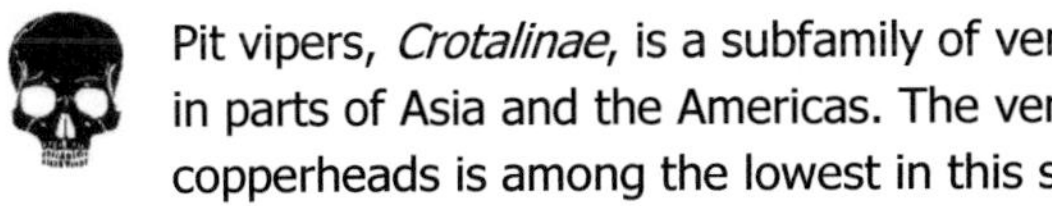

Pit vipers, *Crotalinae*, is a subfamily of venomous vipers found in parts of Asia and the Americas. The venomous strength of copperheads is among the lowest in this subfamily, but a bite should still be taken seriously. Cottonmouths have a slightly stronger venom and the Timber rattlesnake's venom is significantly stronger. Copperheads often use warning bites on humans, just as many other vipers sometimes do. The snake bites but injects little or no venom. The snake is sparing with its venom, but the message has still got through. Delivering dry bites is a good thing for both the snakes and for us humans.

Typical symptoms after a bite are extreme pain, tingling, throbbing, swelling and severe nausea. Damage can occur to muscle and bone tissue, especially when the bite occurs in the outer extremities such as hands and feet. Although this is not an extremely venomous species, bites from a venomous snake should always be taken very seriously. Medical attention should be sought immediately as an allergic reaction or secondary infection is always possible. Deaths do occur after a bite from a copperhead, but it is quite rare.

Copperhead venom has been shown to contain a protein, contortrostatin, which stops the growth of cancer cells in mice. Further studies have been carried out and the results suggest that contortrostatin causes apoptosis in breast cancer cells, inhibiting tumor growth and metastasis. Apoptosis, or programmed cell death, is a way for cells in organisms to die in a controlled manner without harming their surroundings, as opposed to the more chaotic necrosis that can occur after a venomous snake bite.

When the Eastern copperhead feels threatened, it quickly vibrates the tip of its tail. The species does not have a rattle and relies on the substrate, such as dry leaves, to help create sound. This way of warning does not last as long as when a rattlesnake vibrates its rattle. Either the copperhead leaves the site or the snake strikes. If snakes have the opportunity to do so, they always prefer to leave when they encounter humans. Snakes have nothing to gain from a fight with humans, besides the chance to survive.

The act of vibrating its tail as a threat before striking is quite common in the snake world. Many American snakes do this, here at home I have Asian snakes that do the same. A good way to remind us to keep our distance and leave the snake alone!
Copperheads are not large snakes, like many other viper species they are usually under a meter in length. The normal range for body length is 50cm up to nearly a meter, but anything over 80cm is unusual. Record lengths have been recorded at just over 130cm. The body is quite stout, and the head is broad and well-marked from the relatively narrow neck. The eyes have the narrow cat-eye-like pupils that most other species in the viper family have. Eastern copperheads have a pale brown to pinkish-brown ground color that darkens along the dorsal line. As the common name of the species suggests, the head may be considered copper colored. The pattern consists of 10-18 bands or spots on the sides. These are in the same color range as the ground color but are slightly darker with even darker edges. Often these markings are divided at the midline and alternate on both sides of the body, with some specimens having markings that are continuous across the dorsal line. It is also common for snakes to have darker spots in these markings. The belly is the same color as the ground color, or lighter. Along the sides of the belly there is a pattern of strongly marked dark spots, and some scales are slightly darker than the ground color.
Juveniles are more greyish when they are born and with increasing age, they change to the coloring of adults. Juveniles are not entirely in grey, they have a light-colored tail with a yellowish tail tip, just like some other viper juveniles. This bright and rather bright yellow tail is used by the snake to attract small prey such as frogs or lizards.
This behavior is called caudal luring, see page 66.

Photo: Rickard Ljunggren. A juvenile *Agkistrodon laticinctus* with the typical yellow or green tail that juvenile copperheads have.

The overwintering and life cycle of the Eastern copperhead is very similar to that of the Timber rattlesnake, as it is for all species that live where the seasons require a period of the year at frost-free depths. However, Eastern copperheads are a bit less tolerant of cold than Timber rattlesnakes. Copperheads come out of overwintering sites later than Timber rattlesnakes and return earlier. Also, on average copperheads give birth 7-14 days later than Timber rattlesnakes. This is a factor in why they cannot handle the previously glaciated portions of northern Pennsylvania and similar parts of other states, but the Timber rattlesnake can. This is also why copperheads are much more common in warmer states like Virginia and the Carolinas.

The Eastern copperhead begins life above ground as early as March/April and they often live above ground until November. The cold of late autumn signals to the Eastern copperhead that it is time to return to the overwintering site, and both the biological rhythm and the warmth of spring set the end date for overwintering. It also happens that Eastern copperheads comes out of the den to bask in the middle of winter, weather permitting. Otherwise, this species follows the same pattern as the Timber rattlesnake in terms of summer and winter location.

Copperheads reach sexual maturity at around four years of age and mate in late spring or early autumn. Females can retain sperm (often from several different males) and postpone fertilization of their eggs for longer periods, but ovulation and fertilization usually occur in spring. Males fight for the right to mate with females and males that lose a fight will sometimes not challenge another male again. Females may also fight any mates, and she will not mate with a defeated male. The results of these mating fights are an important part of natural selection, with the stronger individuals passing on their genes. This mating ritual is not unique to copperheads, it happens in many snake species.

In autumn, females give birth to anything from a single offspring to more than fifteen young, but the usual number is four to seven. Larger females give birth to a greater number of offspring. Sometimes the female will produce offspring for several years in a row, but they may also skip a year or two. It is commonly said that the young are "born alive", copperheads are ovoviviparous just like Timber rattlesnakes and cottonmouths. The offspring develop inside eggs which remain in the mother's body and hatch therein, then they emerge covered by a thin membrane from which they quickly emerge. The development of this type of live birth reduces the exposure of the egg-encapsulated embryos to both predators and the wrong environment.

Virgin birth can occur through a process called parthenogenesis, a form of asexual reproduction where an unfertilized egg can develop to maturity. This ability is more common in invertebrates. The same phenomenon is found in other reptiles, but is quite rare except in the Mourning gecko, *Lepidodactylus lugubris.*

The young copperheads are up to nearly 20cm long and, like other venomous snakes, are born with both fangs and venom glands. The young snakes can take prey in the weeks remaining before they must enter their first hibernation. Not all newborns succeed in taking their first prey before hibernation, but this does not reduce their chance of surviving the coming winter. At around four years of age they are considered adults, by which time they have reached the size of an adult and are ready to mate themselves. Life expectancy can be up to 18 years, longer in captivity where records are up to 30 years old, but it is not uncommon for Eastern copperheads in the wild to live no longer than 6-8 years.

As its common name suggests, the Eastern copperhead of this chapter occurs in the eastern parts of the USA, but the species has a wide range, occurring in the southern parts of the country as far west as Texas and even into Mexico. It is absent from the southernmost parts of Texas and only occurs in the northern part of Florida. The range is quite like that of the Timber rattlesnake, but Eastern copperheads do not occur as far north, bordering the southern parts of New York State. In Washington, D.C., it is the only venomous snake found.

Photo: Phil Dunning. Newborn Eastern copperhead. In many snake species, the snakes gradually change color, becoming brighter and changing to the color they will have as adults.

Within its range, it is found in a wide variety of habitats, but in much of the area the species favors deciduous and mixed woodland. Eastern copperheads are often associated with rocky outcrops and ledges, but are also found in low-lying, marshy areas. In the states surrounding the Gulf of Mexico, Eastern copperheads are also found in coniferous forests, and in the Chihuahuan desert of western Texas and northern Mexico, they occur in beach habitats. Eastern copperheads are also found in abandoned and dilapidated buildings, construction sites and suburban areas. This very fact probably contributes to the fact that the species is responsible for the most venomous snake bites per year in the US. According to statistics, somewhere between just under 3,000 and up to over 4,000 people are bitten each year by copperheads [14],

but since the species has a milder venom than some other venomous snakes and also delivers dry bites, we probably have a dark figure as not all those bitten seek or need medical attention.
Among the invertebrates, cicadas, caterpillars, millipedes, spiders, beetles, dragonflies, grasshoppers and crickets may be part of the diet. Larger prey may include other snakes, lizards, salamanders, small turtles, frogs, small birds, bats, voles, mice, moles, rats, squirrels, chipmunks and rabbits and even the carcasses of these species. There is a wide variety of prey, but the most common prey is a mouse or other small rodent. Gravid females usually fast, as with many other snake species. Like many other species of viper, the Eastern copperhead hunts mainly by ambush. They lie motionless in a suitable spot, waiting for a prey animal to come too close. The exception is when the snake is hunting insects, when they hunt actively instead. Younger specimens also hunt by attracting prey with the technique you just have read about.
Sight, smell and heat detection are used to locate prey. When Eastern copperheads take smaller prey or birds, they do not let go after biting but keep the prey in their mouths until it dies. When they bite larger prey, they let go immediately after biting, so as not to risk injury when the prey tries to defend itself or escape. Once the snake has bitten its prey and let go, the venom is allowed to do its work while the snake searches for its newly bitten prey. After the bite, smell and taste become the primary means of tracking down the prospective meal. Envenomated prey is easier for snakes to locate and find than other potential prey of the same species that happen to be nearby. The scent of the snake's own venom in the prey helps to guide the snake in the right direction and they can distinguish their prey from the scent trails of its species relatives.
Owls and hawks are the most common predators of copperheads, but owls mainly take younger and smaller specimens. The opossum is one of the most successful predators and is sometimes used to keep copperhead populations down. The species is immune to copperhead venom, which means that the snake has virtually no way to defend itself. Bullfrogs hunt at night and will also take juvenile copperheads. Other known predators are bears, weasels, raccoons and alligators. There are several snake species among the predators. Black racer, king snakes, Eastern indigo snake and cottonmouths eat copperheads.

10. Northern cottonmouth

Family	*Viperidae*	Vipers
Genus	*Agkistrodon*	American moccasins
Species	*Agkistrodon piscivorus*	Northern cottonmouth

Photo: Dr. Edward J. Wozniak. A Northern cottonmouth displaying its famous white mouth.

The genus name *Agkistrodon* is the same as the Eastern copperhead in the previous chapter. The species name, *piscivorus*, comes from the Latin *piscis* (fish) and *voro* (I eat greedily). This gives the species the name "fishhook-teethed greedy fish-eater", which in my opinion does not sound as good as the English common names. The species has many names in English including Northern cottonmouth, Cottonmouth, Water moccasin, Swamp moccasin and Black moccasin. These are the most frequently used of over 60 different common names. This jungle of names can sometimes cause confusion, which is why many herpetologists tend to use the scientific names. To add to the confusion, copperheads are sometimes called Upland moccasins. Sometimes it can also be the case that a common name is used for several different species. An example is the Eastern copperhead in the previous chapter. In English it is often called Copperhead, but Copperhead is also the common name of three species of venomous snakes in Australia**,** in the genus *Austrelaps* (family *Elapidae*). They are venomous elapid snakes that is native to the southern and eastern part of the Australian continent. They are commonly called copperheads or Australian copperheads.
The scientific name is only ever one of which is current, there are never different species sharing a scientific name and the name is international. I would think that Cottonmouth is the name most commonly used in the USA. The name Cottonmouth refers to the white mouth that the species shows when it feels threatened, see picture on the previous page. The same phenomenon is found in the Black mamba, *Dendroaspis Polylepis*, in sub-Saharan Africa, where the snake shows its jet-black mouth. At home, I have the Green cat snake, *Boiga cyanea,* in a terrarium. The Green cat snake also shows a jet-black mouth when it feels threatened, but unlike the Black mamba, the species has a rather weak venom.
Water and Swamp in the common names refer to the habitat in which the species is often found, moccasin comes from the genus name. The genus is called moccasin which is the Algonquian word for shoe. Algonquian is the collective name for a group of related languages spoken by indigenous peoples in the United States and Canada. The origin of this common name is unknown, the first time it appears in writing is in 1765. It is guessed that its use relates to the color and appearance of snakes or to the silence with which they move.

Photo: Pixabay. Cottonmouths are the most aquatic species of the genus *Agkistrodon,* and they are associated with water courses such as streams, marshes, swamps and the shores of ponds and lakes. These species are also considered to be among the most aquatic of all species in the *Viperidae* family.

Photo: WikimediaImages, Pixabay. Moccasins made by the Arapaho, a North American indigenous people who lived on the prairies.

A threatened Northern cottonmouth vibrates its tail, just like many other vipers and other American snakes do. More typical for the species is the display of its white mouth, the mouth with cotton in it. If you see a snake in the USA that throws its head back and shows a white mouth, you can be sure it is a cottonmouth. Cottonmouths often hiss loudly when threatened, just as many other snake species can do. Other defensive reactions include flattening the body to look larger and emitting a strong-smelling secretion from the anal glands located at the base of the tail. This secretion can be ejected in thin jets if the snake is sufficiently agitated or if it is restrained. The odor has been likened to that of a goat or smelly seaweed.

Northern cottonmouths have a stronger venom than the Eastern copperhead of the previous chapter has. The fangs are hollow like a needle and curved backwards, just like the fangs of copperheads and rattlesnakes. All vipers are solenoglyph*, with long fangs on the front of the upper jaw that resemble hypodermic needles. The ability to fold their fangs back towards the palate means that vipers can have longer fangs than other venomous snakes. Snakes such as cobras have fangs that mainly are in a fixed position with a restricted degree of rotation. With the help of individual muscles, vipers can articulate their fangs together or individually. During a viper strike, the mouth can be opened to almost 180°, the upper jaw rotates forward, and the snake rotates the fangs forward as late as possible so that they are not damaged (as they are rather fragile). The long fangs of vipers means that the fangs penetrate rather deeply.

*Having solenoglyphic teeth means that the fangs that transport the venom into the prey are hollow. The fang simply looks like a needle.

The jaws of vipers close when in contact with the prey/threat and the muscular sheaths encapsulating the venom glands contract and pushing the venom down the venom duct and through the fangs as they penetrate the target. This strike is extremely fast and in the case of defensive strikes more than one bite can occur. In defensive attacks, the vipers sometimes deliver dry bites to save venom. In addition, to having the ability to deliver dry bites in defense, the vipers can also dispense venom when biting their prey. Larger quantity for larger prey and smaller quantity for smaller prey. It is even the case that vipers can regulate the amount of venom according to the species of prey or predator (threat)! Nature is fantastic!

Localized swelling occurs around the bite, when bitten by a Northern cottonmouth. This is to be expected after a venomous snake bite regardless of the species that bit. The pain after a bite is generally more severe than after a copperhead bite, but less than after the bite from many rattlesnake species. The venom from cottonmouths is tissue-destructive and can also cause hemorrhagic disorders, bites can leave severe scars and in rare cases require amputation. Deaths are rare, effective serum is available that is produced from the venom of cottonmouths, Mojave rattlesnake (*Crotalus scutulatus*), Eastern diamondback rattlesnake (*Crotalus adamanteus*) and Western diamondback rattlesnake (*Crotalus atrox*). Bites from cottonmouths are not entirely uncommon from lower Mississippi and the other states around the Gulf of Mexico. Many people encounter these snakes in the wild, but just like other snake species, cottonmouths choose to avoid humans or ask us to keep our distance by threatening us. Venom is more useful in a prey animal than in humans; venomous snakes use venom to hunt for food rather than for defense.

The Northern cottonmouth is the largest species in the genus *Agkistrodon*. It usually grows to about 80-90 cm and is quite heavily built. As with many other species, there are a lot of tales circulating about size records, it is not only fishermen who find it difficult to give the correct size of animals encountered in the wild. The largest confirmed specimen in the wild was 188 cm long. This cottonmouth was captured and given to the Philadelphia Zoological Gardens. Unfortunately, the snake was injured during capture and died a few days later. A specimen of this size weighs over 4kg.

The broad head is distinct from the neck and the snout is blunt in profile with the edge of the top of the head extending forwards slightly further than the mouth. The scales are keeled, as in copperheads, rattlesnakes and many other snakes. Keeled scales have a ridge in the center which makes them feel rough to touch. There are theories about the function of keeled scales, that because the keeles reduce the shininess of the reptile, it makes it easier for the animals to hide, which may be an evolutionary advantage.
In some vipers, particularly those of the genus *Echis*, the scales are not only keeled, but the keel has small serrations. These snakes use this to warn and threaten. By wriggling against their own body, they make a hissing sound through the friction that occurs. This method of making sound is called stridulation.

The majority of Northern cottonmouths are brown, blackish-brown or even completely black, except for the head and facial markings. These snakes may have a pattern consisting of a yellowish-green, brown, greyish-brown or blackish ground color overlaid with a series of dark brown to almost black cross bands. These cross bands, which usually have black edges, are sometimes interrupted along the back of the snake. These crossbands are visibly lighter in the middle and there they almost match the ground color. The crossbands often contain irregular dark markings and extend far down the sides of the snake.

Juveniles and subadults generally have a lighter ground color and a more contrasting color pattern, with dark crossbands on a lighter ground color. The juvenile ground color is yellowish brown, brown or reddish brown. The banding pattern gradually fades as the Water moccasins become darker with age, older individuals are almost uniformly colored olive-brown, grey-brown or black. The belly of the water moccasin is white, yellowish-white or brown and marked with dark spots, the belly becomes darker at the back. The underside of the head is usually white, cream-colored or brown. The tip of the tail is yellowish in juveniles, becoming greenish-yellow or greenish in subadults and then black in adults. In some juveniles, the banding pattern may also be visible on the tail.

The water snakes of the genus *Nerodia* are often mistaken for cottonmouths and are sometimes killed for this reason. These snakes are large-headed and roughly have the same color range as the cottonmouths. However, water snakes have round pupils, lack the typical nasal pit and never vibrate their tails when threatened. Another difference is that a water snake usually flees from an encounter with a human, cottonmouths often lie still and show its white mouth. Water snakes are completely harmless to us humans, they have no venom. Both water snakes and cottonmouths spend a large part of their lives in or near water.

Northern cottonmouths are found in the eastern United States, which is perhaps not surprising when you have gotten this far into the book. The species can be found in Alabama, Arkansas, Georgia, Illinois, Indiana, Kentucky, Louisiana, Mississippi, Missouri, North Carolina, Oklahoma, South Carolina, Tennessee, Texas and Virginia. In general, Eastern copperheads and the Timber rattlesnake are found in the same area, but there are no cottonmouths in the northern parts of the range of the other two species. However, cottonmouths (The Florida cottonmouth, *Agkistrodon conanti.*) are found further south than Timber rattlesnakes and copperheads.

In the northern parts of their range, Northern cottonmouths hibernate. In many respects their life cycle is similar to that of the Eastern copperhead. Hibernation is not very long; they are often reasonably active until the first heavy frost and hibernation ends quite early in the year. In the southern parts of its range, hibernation may be short or omitted altogether. In fairly large parts of their range, they can be found all year round, even on sunny days in wintertime.

Like copperheads, cottonmouths mate in both spring and fall, but mating usually occurs in April or May and the offspring are born in August or September. Females give birth to an average of 5 to 9 offsprings but can have as many as 20 or as few as 1.
(For hibernation, sexual maturity, mating and mating rituals, birth, and virgin birth, see page 158, Northern cottonmouths and Eastern copperheads function in much the same way.)

A newborn Northern cottonmouth is slightly larger than a newborn copperhead, from closer to 20 cm up to 30 cm. Female cottonmouths can reach sexual maturity slightly earlier, as early as three years of age, if the weather and food availability are favorable. Northern cottonmouths often watch over their newborns, more on this on page 169.

Although cottonmouths and copperheads are closely related and have similar ranges, their habitats differ somewhat. Cottonmouths are the most aquatic species of the genus *Agkistrodon*. They spend much of their life near water bodies such as streams, irrigation ditches, marshes, swamps and the shores of ponds and lakes. The species can be found in brackish water and sometimes in salt water, but this is somewhat rarer. They are not usually found in deep, cold water, as they are, after all, hypothermic animals. Larger specimens can be found further away (up to two kilometers) from water than smaller specimens. In some places, cottonmouths have adapted to forest and prairie environments.
Cottonmouths are terrestrial and also spend some of their time in or on water. They like to bask on logs, rocks or lower branches. If you see a cottonmouth a few meters up in the branches or higher, it is probably a water snake (*Nerodia)* you are looking at. It also happens that cottonmouths are curled up and floating on the water surface. You can find clips showing this on YouTube.

The species can be active both day and night. Food is mainly hunted after dark during the warmer parts of the active season. They hunt both by ambush and by actively seeking out prey. Younger specimens also hunt by attracting prey, as described on page 66. Cottonmouths are also opportunists; there are probably very few animals of the right size that could not become a meal. Their diet includes a variety of aquatic and terrestrial animals. They eat amphibians, lizards, snakes (including cottonmouths), small turtles, small alligators, small mammals, birds and fish. The carcasses of these animals may also be ingested, as well as fish scraps and slaughterhouse waste. Because they eat dead things, especially in dried up pools or off the road, secondary infections from cottonmouth bites are also an issue aside from the venom.

Small prey is held on to after biting, but larger prey that could harm the snake is released immediately and the snake waits, letting the venom do its work before approaching again.

There are documented cases where cottonmouths (The Florida cottonmouth, *Agkistrodon conanti.*) have eaten smaller specimens of the invasive Burmese python. The Burmese python, *Python bivittatus,* is a species from Southeast Asia and one of the five largest snake species in the world. Following the escape (and in some cases deliberate releases) of Burmese pythons from captivity in large numbers, the species has become established in Florida. The Burmese python is most often found in or near water bodies.

Cottonmouths can become food for countless animals. Among the reptiles that eat them are snapping turtles, alligators and other snakes. In addition to cannibalism, king snakes and black racers also eat cottonmouths. Among birds of prey, it is mainly great horned owls, eagles, falcons, hawks, herons and cranes that cottonmouths need to watch out for.

As I have written before, the species gives birth to live young, just like almost all vipers. The newborn snakes stay with the female for up to two weeks, usually until the first molt [15]. Gravid females often shelter together in pairs or trios. Usually, they give birth in a well-protected burrow near water where they share the work of watching over their common offspring during the vulnerable first week. The bottom picture on the next page shows two females and a week-old offspring that has molt for the first time. As you can see in the picture, the female on the right will soon molt.

Although they do not feed their newborns, they are extremely watchful and offer protection from predators, especially other snakes. There are also documented cases of offspring staying longer with the mother.

Photo: Dr. Edward J. Wozniak. Photo taken in Tarrant County (Arlington) Texas. A younger Northern cottonmouth that still has the juvenile, slightly sharper pattern.

Photo: Frederick S Boyce. Picture taken near the coast of North Carolina. Female Northern cottonmouths with a juvenile.

11. Death and dying

I actually finished this book in November 2024, and it was sent for proofreading. When I went through my text and made the suggested corrections and changes, I felt that the book was missing something. Hence the addition of this eleventh chapter, with an additional focus on the often-present death.
I must also admit that it is a subject that I find hard to stop writing about, it is extremely exciting to explore. Reading about lives that have ended tragically, reading death certificates and old news articles and, in some cases, even chatting with relatives alive today is one of the most interesting things I have ever done.

*

For Pentecostals, spiritual and physical healing serves as a reminder and testimony to Christ's future return, when his people will be completely delivered from all the consequences of the fall.
Pentecostals believe that prayer and faith are central in receiving healing, but everyone does not receive Gods healing through prayers. It is God in his wisdom who either grants or withholds healing.
Common reasons that are given as answers to the question as to why all are not healed are:
God chose to call the sick person home.
God teaches through suffering.
Healing is not always immediate.
Lack of faith on the part of the person needing healing.
Personal sin in one's life (this does not mean that all illness is caused by personal sin).

Everyone can pray for one's own healing and for the healing of others, no special gift or clerical status is necessary.
The sick person expresses their faith by calling for the elders of the church who pray over and anoint the sick with olive oil. The oil is a symbol of the Holy Spirit. Besides prayer, there are other ways in which Pentecostals believe healing can be received. One way is based on Mark 16:17–18 and involves believers laying hands on the sick. This is done in imitation of Jesus who often healed in this manner.

During the initial decades of the movement, Pentecostals thought it was sinful to take medicine or receive care from doctors. Over time, Pentecostals moderated their views concerning medicine and doctor visits, but a minority of Pentecostal churches continues to rely exclusively on prayer and divine healing.

In terms of the approach to seeking medical care, my impression is that the churches that deal with venomous snakes have not fully made the same journey that much of the rest of the Pentecostal movement has made. Big changes have taken place, I think medical care for venomous snake bites is the final dividing line. A few decades ago, it was unusual to seek medical attention after a bite, nowadays it's probably becoming more and more unusual not to seek treatment.

I asked Cody Coots to comment on the text you've just read, regarding whether or not to seek care. I received his reply the very next day:

"I believe in laying hands on the sick and then recovering. I've also experienced a miracle of healing immediately and seen a lot too. As for the medical attention after a snake bite it's an individual thing. Some people don't believe in going and some do. I don't judge either way. When I went for that bite in the head, I was treated really bad by the doctor in Middlesboro, Kentucky and it put a bad taste in my mouth. The last time I got bit with a Western diamondback that was the biggest factor why I did not want to go to the doctor. Thankfully the Lord helped me with the pain and the sickness that I did not have to go."

"One time, he said he believed if you didn't handle rattlesnakes you were going to hell. That seemed to me to make for a very small heaven."
Author Dennis Covington on a conversation with *Bill Pelfrey.

Personally, I'm not entirely sure there is a heaven to go to, and thus I also doubt whether there really is a hell to go to if you haven't handled rattlesnakes or otherwise earned eternal damnation.
Bill's view is of course extreme. Those who handle rattlesnakes for religious purposes often talk about getting closer to God, but the view that the rest of us go to hell is unusual.

*Billy Ray Pelfrey, 80, passed away Friday, April 13, 2018, at his residence. He was an evangelist with *the Holiness Church of God* in Virginia. According to legend, his first wife Anna Ruth Pelfrey, had died twice after being bitten. Both times she was revived through prayer. She died in 1998, but the source of the legend is older.

*

The following text lists some notable deaths that, to me, stand out a bit. Some of them were well known for their practice of venomous snakes, some of them are rather or entirely unknown.

On page 24 I write about the first confirmed death (from a venomous snake bite) within the Pentecostal Church that I have found. I have also found information that the earliest death is from 1906. In other words, this is before George Went Hensley's time as a snake handling pastor. Hensley was not the first, but the practice of snake handling did not gain national attention until he entered the scene in 1910.
I can't quite get the timeline right, though, as the bitten one is supposed to have attended Hensley's sermon.
"The first life claimed by snake-handling was a young man in Bartow, Florida. In May 1906, he was bitten by a snake during one of Hensley's sermons and became ill. Though Hensley predicted the man would miraculously recover, the man died soon after as a result of his carelessness in handling the snake during the sermon. Bartow, in turn, passed a law banning snake handling, and Hensley left the area shortly thereafter."

A (pretty good) guess on my part is that there is a digit wrong in the year stated as I also found this in a newspaper called *"The Sentinel"*, printed May 5, 1936:

"FAITH" SNAKE-BITE FATAL
Bartow, Fla., May 4. -Alfred Weaver, 35-year-old itinerant bitten by a rattlesnake during a "faith" demonstration at a revival service here last night, died today refusing medical treatment. He had declined to permit physicians to aid him even after the hand on which he was bitten turned purple and his arm was swollen to the shoulder. During a revival meeting conducted by George Hensely, who frequently has handled reptiles in the course of his demonstrations, Weaver picked the snake from a box. It lashed out and struck him twice.

*

In Cleveland, Tennessee, eighteen-year-old Harry Skelton was bitten and died in the year 1945. Five days after his death, Walter Henry was handling the same serpent that killed Skelton, and he was bitten and died. Henry's brother-in-law, Hobert Williford, handled a snake at Henry's funeral where he was bitten and he also died.

*

On July 23, 1945, *the New York Times* featured a story titled, *"Snake Bites Man and Dies"*, about a snake handler named Luther Morrow in Grasshopper, Tennessee. On July 22, Morrow was bitten by a large rattlesnake in a church service. *"The following morning, the snake died."* David L. Kimbrough, researcher of the snake-handling faith, claims he has seen many rattlesnakes die in the hands of believers. He said that some die from over-handling and many die within a month of being placed in captivity.

*

In September 1945, at a Virginia meeting with snake handling in Wise County, Anna Kirk got bitten. The 25-year-old* pregnant wife of Reverend Harvey O. Kirk was bitten three times on the wrist after patting the head of a snake and waiving her arms over it.
Her hand became swollen and turned black. Although she was pregnant, she refused to seek medical care. Three days after the bite, she went into labor, after six months of pregnancy. The baby died moments after childbirth, which was conducted without a physician present. Shortly afterwards Anna also died.
Samples of Anna's blood revealed that she, and likely to her newborn, died from the venom of the snake.
Anna's husband was arrested for murder for handing her the snake, and ultimately pleaded guilty to manslaughter and accepted a sentence of three months in jail.

* Another source indicates that her age was 26.

Pastor Harvey O. Kirk was found guilty of involuntary manslaughter in the death of his wife and unborn child. Judge George Morton sentenced Harvey Kirk to two years in prison, but Kirk's attorney made a motion to set aside the verdict, which was denied at a hearing on August 29, 1946. That decision was appealed and in October 1947, the Virginia Supreme Court overturned it. A new trial was held in August 1948, and Kirk agreed to plead guilty to manslaughter in return for a three-month sentence. He asked the court for a few days to wrap up his business and Judge Morton granted it. When the day came for him to return to court, Kirk never showed up. He apparently had a change of heart and had skipped town. In November 1948, he was located in a tourist camp near Winter Haven, Florida, and captured by Polk County authorities and returned to Wise County.

*

On August 29, 1955, Anna Marie Yost, age 46, was bitten on the arm while handling a rattlesnake during a religious service in Savannah, Tennessee. Her brother, Mansel Covington, a well-known snake handler, was bitten on both hands during the same service but survived.

Anna Marie and her brothers Allen, George and Mansel came from a family of snake handlers. The three brothers had already been arrested after they brought a snake into the Bumpass Creek worship service and thus created some turmoil and chaos. The three made bail and were free in time to attend a two-week snake-handling revival in a private home in the Walnut Grove community of Hardin County. According to Allen, Anna Marie was bitten by a rattlesnake*, "a pretty good sized one, he had six rattles.* Anna Marie Yost received no medical treatment, but was instead taken home, *"after we couldn't prevail on the Lord to do anything."* She died the next day.
When Anna Marie died, George contacted their sister Edna who was working night shift at a hospital in Louisville. She didn't see any reason to rush back to Savannah, since Anna Marie was already dead, but George insisted that she leave work right then and drive all night to get to Savannah. George said he and his brothers were going to raise Anna Marie from the dead through prayer, and since Edna was a nurse, he wanted her to be there to check his sister's vital signs.

Her brother Mansel, who also got bitten, refused medical care. The court intervened, and he was taken to Savannah for treatment. Because of that ruling, Mansel survived.
Mansel, as well as Thurston Frayser (the minister) and his wife, were arrested and charged with violating Tennessee's snake-handling law. According to Hardin County General Sessions Judge John Caldwell Covington and Frayser were to be bound over to a grand jury on the charge of felonious homicide.
The day after his arrest Mansel Covington told the court that his sister died *"because her faith was failing. She got scared of the snakes and that's what killed her."* For his part, Hardin County Coroner Judge John Caldwell offered his opinion: *"What we've got is a bunch of poor, ignorant people up here, and we're trying to make them see what's right for their own good."*

Anna Marie's five or six children (the number differs between sources) grew up without their mother, and her brothers had to live with the knowledge that they had some part in her death, even though ultimately the decision that cost Anna Marie her life was hers to make.

In the newspaper *The Tri-Cities Daily* on August 30, one could read about the incident. An odd thing is that when writing about Mansel, the newspaper states his weight at 300 pounds, something I find difficult to see as relevant in this context.

*

On September 28, 1961, Columbia Gay Hagerman, a 22-year-old divorced housewife died after getting bitten by a Timber rattlesnake. She was bitten on the right thumb during her very first snake handling at a church service in Jolo, West Virginia. She declined medical assistance and died at her parents' home.

Her stepfather, Bob Elkins was a coal miner and her mother, Barbara Elkins was a home maker who raised her six children from a previous marriage. The Elkins formed *the Jolo Pentecostal church* in 1956 where 50-100 members attended every week. Her parents had previously been bitten several times by copperheads and rattlesnakes, recovering each time without seeking treatment. Her older brother, Dewey Chafin states that he during a lifetime got bitten by various venomous snakes over 100 times. According to her death certificate, she had been bitten five days earlier, and she had all the time in the world to question her decision not to seek medical attention. Under *"Name of hospital or institution"* they have filled in *"none"*, under "*Immediate cause"* it says, *"snake bite on right thumb from poisonous snake."* Finally, the coroner has filled in that it was an accident.
I happened to find a relative of Columbia's who is alive today, who also tells me of another relative who died after handling venomous snakes during church services*. Tanya, the relative that I found online, tells me her mother-in-law remembers the incident when Columbia was bitten. The mother-in-law was there when Columbia got her fatal bite and at Columbia's parents' house before she died. She was only a child, but still remembers.

*Three years prior Columbia's death, Hazel Robinson Payne died. She was taken to the home of a member of *the Jesus Only Church,* where she, her son and her daughter repeatedly refused to let a doctor treat her bite. Her husband, who came home after being notified about the incident, said he would abide by his wife's wishes in the matter. She died eight hours after the bite. Another member of the congregation told authorities that Hazel arrived late for the service and immediately went to the front of the church where a number of rattlesnakes had been released. She picked up one of the larger ones and *"fondled it like a kitten",* the eyewitness said. The snake struck at her as she turned to hand it over to a male congregation member.

*

On Tuesday, February 21, 1967, *the Owensboro Inquirer* reported that a 24-year-old man had been bitten by a rattlesnake during a service at *the Free Holiness Church.* James Saylor, as the man was called, was taken to hospital the following day but was dead on arrival.
William D. Saylor, age 39, was minister in that church. William and two others went on trial for violation of a Kentucky law which bans use of snakes in religious services. The trial stemmed from the death of Williams nephew, James.
8th of March 1995, Kale Saylor, 77, a longtime Pentecostal preacher, died after he was bitten on the right hand by a rattlesnake at a church in Crockett where he was preaching. Saylor's first wife, Jean, died in December 1967 after being bitten by a rattlesnake in Bell County.

The more I look into old events and deaths, the more I am convinced that snake handling takes place in rather small parishes and that the members of a parish often consist of a small number of families. People are born into the faith and some die for their faith.

*

In *the Rushville Republican* on September 25, 1972, in a small notice on page 8, one could read about another fatal snake handling.
Beluah Bucklen, 59 years old, had been bitten twice by a rattlesnake and had died the day before. The bites occurred during a service at *the Jesus Pentecostal Church* in Putnam County on September 16. Her

husband had convinced her to seek medical attention when she became violently ill the day after being bitten, but despite treatment she died eight days after the service. Roscoe Bucklen, her husband who was sitting in his car outside the church when his wife was bitten, said, *"I've seen that snake before. It's as big around as your arm. It hit her twice between the thumb and forefinger on the left hand."*

In *the Springfield New-Leader*, May 21, 1976, page 6, the newspaper told the story about Curtis Mount, who got bitten twice on the right hand by a really big water moccasin during a service at a church in Mingo County. Curtis Mount sought no medical attention before he died three days after the bite. The coroner said that the fang marks were an inch apart, almost twice the distance of marks from an average bite.

"The right arm was swollen twice the normal size from the hand to the shoulder and was discolored" according to Ernie Ritchie, the coroner. He also told the newspaper that the swelling had entered into the chest area, along with discoloration.

The bite produced blood poisoning and led to paralysis of the diaphragm and coronary muscles. Curtis Mount was also drinking strychnine during the service when he was bitten.

The verses from chapter 16 of the Gospel of Mark in the New Testament, play a major role for the members of *The Church of the First Born*. On June 16, 1982, nine-year-old Jason Dean Lockhart died in Enid, Oklahoma. Jason Lockhart's parents, Palmer and Patsy Lockhart, refused to take Jason to the doctor after complaints of not feeling well for over five days because it was against their beliefs. The parents believed that Jason was suffering from food poisoning, but on the fifth day Jason began to worsen as he began sweating profusely and complaining of stomach cramps.

Palmer Lockhart stated the last day Jason went to the restroom around five and then went back to the couch where he later died.

Enid Police Department Detective, Michael Danhy, later testified that when he arrived at the residence that Palmer Lockhart said he did not call a doctor because it was not worth his son's soul burning in hell.

When Palmer Dean and Patsy Lockhart were told by an Enid police detective that their son died because of complications stemming from a ruptured appendix, they started crying.
Jack Robinson, an elder of *The Church of the First Born*, prayed over Jason while he was ill and later testified that if Jason would have received medical attention from a doctor that he would have been forgiven. Dr. A.J. Chapman of the Oklahoma State Medical Examiner performed an autopsy on Lockhart and found that he had a ruptured appendix. When Palmer Dean and Patsy Lockhart were told by an Enid police detective that their son died because of complications stemming from a ruptured appendix, they started crying.

"Mr. Lockhart stated to me that, if he had it over to do again, he would not call a doctor, he said that it wasn't worth his or his child's soul burning in hell."
Words from Enid Police Sgt. Michael Danhy.

Dr. Chapman testified in court that Jason had gone through *"excruciating pain"* before he had died. The District Court of Garfield County charged Palmer and Patsy for manslaughter in the first degree. Judge Bussey stated in his opinion that Palmer and Patsy Lockhart were being charged on the basis of *"willfully and wrongfully, without lawful excuse, therefore, omit to furnish necessary medical attention for their minor child, directly causing his death".* On December 15, 1982, the Garfield County district court jury acquitted Palmer and Patsy of the first-degree manslaughter charges. Harold Moler, jury spokesman, stated that the jury was not happy about their decision, but did so because they were not guilty under the statute.

*

February 19, 1986, Shirley McLeary from Ohio died after getting multiple bites from an Eastern diamond rattlesnake seven hours earlier. The incident happened during a service for her uncle´s funeral in Baxter, Kentucky. She was bitten *"at least three times"* when she handled the Eastern Diamondback rattlesnake during a wake at *the Church of Jesus Christ* in the mountains of Harlan County.

She did not seek medical treatment, as the other church members were praying to perform a "faith-healing attempt" on her. The prayers went on for all seven hours.
Kentucky State Police spokesman Bill Riley was interviewed.
"They pray over them and lay hands over them and practice faith healing. They are supposed to be filled with the spirit of God and not supposed to die. If a person bitten by a snake declines medical help, I don't know what we can do", said Riley, adding it had not been determined whether McLeary herself did not ask for treatment.

*

Arnold Loveless died 9th of April 1990. He was bitten in the jaw by a Timber rattlesnake during a service at *the Church of Jesus Christ*
in Cartersville, Georgia.
After getting bitten he began to pray, and fellow worshipers laid hands on him in attempt to heal the bite. Witnesses said emergency medical workers were not notified until Loveless became violently ill. He went to an intensive care unit, but his life couldn't be saved.
The snake venom killed Loveless relatively quickly because it entered his body through the salivary glands – *"the quickest way into the system",* according to Bill Conyers, a nursing supervisor at Humana Hospital where Loveless died.
The minister Carl Porter said, when interviewed, that it *was "the man's time to go".* The minister also told the newspapers that *"the Lord wouldn't have let it happen if it wasn't his time to go, it wasn't no judgment on him, he was a fine man".* He told the newspapers that the church use snakes that are native to the area, including rattlesnakes, copperheads and cottonmouths. He added that *"we've had a cobra and some others I didn't know what they were".*
The Timber rattlesnake who bit Loveless was taken to the hospital so that doctors could determine the correct remedy, then it was taken to an animal shelter.
Six days after Arnold Lee Loveless, 48, was fatally bitten by a rattlesnake and two days after his funeral in the same wood-paneled sanctuary, it was worship as usual in t*he Church of the Lord Jesus Christ* between Cartersville and Rome.
The minister Carl Porter warned the three-dozen people in the pews, *"We've got some serpents up here. They'll bite you, and they will kill*

you. If you get one of them out, that's between you and the Lord."
When asked about Loveless' death, Porter said he believes each person's fate is predestined from the beginning of the world.
"I don't believe anybody can die until God gets ready for them."
The Timber rattlesnake who bit Arnold Loveless was returned to the church at the request of a member of the congregation.

*

The Associated Press, July 15, 1991:
A man whose stepfather appealed a state ban on snake handling to the U.S. Supreme Court has died from a snakebite.*
Jimmy Ray Williams Jr., 28, of Hot Springs, N.C., was bitten on the right arm by a 3-foot black Timber rattler Saturday during a service at the House of Prayer in Jesus Name. Authorities said snake handling is classified under Tennessee law as cruelty to animals, punishable by six months in jail.
"People at the house were very close-mouthed about it, even though it's just a small misdemeanor,'" said White Pine Police Chief Jeff Manis.
Williams' father, Jimmy Ray Williams Sr., died in 1973 after drinking strychnine during a worship service. Snakes were handled at his funeral.
Jimmy Ray Williams was bitten two days prior his death.

New York Times 10th of April 1973:
NEWPORT, Tenn., April 9—A preacher and another leader in church snake handling died early yesterday after drinking strychnine at a service that Cocke County officials said was a demonstration of faith.
The dead were the Rev. Jimmy Ray, Williams, 34 years old, assistant pastor of the Holiness Church of God in Jesus Name in Carson Springs and Buford Pack, 30, from Marshall, N.C., the brother of the Rev. Liston Pack, the pastor of the church.
The police said the two men had refused medical attention.

Another church member got bitten by a rattlesnake during the same service in 1973 but survived the bite. Sheriff Bobby Stinson said the authorities had learned of what happened when members of the Newport Rescue Squad were called out late Saturday night and were told that Mr. Pack and Mr. Williams had taken strychnine. B. C.

Ramsey, a deputy, said that Mr. Pack was dead when the squad arrived and the crowd *"refused to let the rescue squad bring Wiliams to the hospital."*

*His stepfather was Liston Pack, pastor in the congregation in Carson Springs where his father drank strychnine and died.

*

Nov. 28, 1991, *Los Angeles Times* had an article about Dewey Chafin and *the Church of the Lord Jesus* in Jolo. Chafin, among some parishioners, were interviewed. One of the parishioners was Ray Johnson who got four lines of text in the article:
"Here is Ray Johnson, who drives 150 miles to the church every weekend from Galax, Va., with his wife, Betty, and their five children. They bunk in a vacant cabin on a nearby mountaintop. Johnson has had two heart attacks. He wears a cap reading "God Said It, I Believe It And That's It."
December 2, 1991, Ray Johnson was bitten twice on his left wrist by a Timber rattlesnake in *the Church of the Lord Jesus* in Jolo. He refused medical treatment and died thirteen hours later.

*

The first known death, from a snake bite during service, this century happened in Tennessee. Darrell Lynn Fee, a resident of Rose Hill, Virginia, was bitten in the chest by a Timber rattlesnake during a service at a church near LaFollette. He did not seek medical attention and did not want local authorities to know he had been bitten. He later died from the snake bite at the age of 45, on August 29, 2000.

*

November 5, 2006. Linda Long, 48 years old, succumbed to the snake´s venom little more than three hours after she was bitten on the right cheek. She was bitten by a yellow Timber rattlesnake when she attended a service at *East London Holiness Church* in Kentucky. Other parishioners at that service quickly took Long to *Marymount Medical Center* in London. Lt. Ed Sizemore of *the Laurel County Sheriff's Office* said friends went with Long to a local hospital Sunday afternoon, before she was transferred to the university hospital. Neighbors of the church told the newspaper the church practices serpent handling. Handling reptiles as part of religious services was (and still is) illegal in Kentucky. Snake handling is a misdemeanor and was then punishable by a $50 to $100 fine. Police said they had not received reports about snake handling at the church.

Her family sued the hospital and several of its employees. According to the lawsuit, on the way to *Marymount Medical Center*, someone in the vehicle called 911 at 7:46 p.m. A dispatcher connected the call to the hospital and the driver asked for an air ambulance to fly Long to Lexington. Hospital employees assured the Long family a helicopter was available. A nurse met Long and those with her in the parking area outside the emergency room. Rather than take Long in right away, the nurse engaged Long and her family *"in a lengthy and time-consuming series of questions"* that went far beyond getting information needed to treat the snakebite, the lawsuit stated.

After being taken into the hospital at 8:09 p.m., Long said she was having trouble breathing and she asked for oxygen. Hospital employees gave her a portable, oscillating fan as they allegedly *"snickered and made derogatory comments"* to other employees and Long's family about the religious beliefs and the circumstances under which she was bitten.
Her blood pressure dropped, her heart rate went up and her neck, face and tongue swelled, and she went into shock. However, a doctor failed to properly treat her and did not put in a tube to help her breathe, according to the lawsuit.

At 8:28 p.m., hospital personnel contacted the air ambulance service. When the helicopter arrived 12 minutes later, the crew asked the doctor to put in a tube to help Long breathe, but the doctor said her airway was not the problem and told the flight crew to quickly get her to *University of Kentucky Medical Center* in Lexington, the suit said. Linda Long's heart stopped on the way. She was pronounced dead at 10:50 p.m. at the *University of Kentucky Medical Center*.

The suit said the failure to quickly and properly treat Long contributed to her death, and that the failure to give proper treatment contributed to the severity of her condition and *"resulted in her ultimate demise"*. The hospital failed to adhere to proper standards of care, according to the lawsuit. The complaint also said the unprofessional comments about Long's religious beliefs were discriminatory and caused her and her family emotional pain and humiliation.

*

"I felt like God would heal him": Mother on trial for death of son, 9, from untreated diabetes.
On May 25, 2012, *the Daily Mail* reported that Susan Grady was charged with second degree manslaughter for the death of her son Aaron on June 5, 2009, from complications related to diabetes. Prosecutors argue that the 43-year-old woman acted negligently by allowing her son's condition to worsen by depriving him of medical attention. She posted the $10,000 bond and was released by noon, according to jail records.
In an audio-recorded interview, Grady told Broken Arrow Police *"I felt like God would heal him, we just believe that prayer works".*

Friday, May 25th, 2012, you could read this on *Newson6*:
A Tulsa County jury reached a verdict after 4 hours of deliberation in the trial of a Broken Arrow mother charged with manslaughter. Susan Grady was found guilty Friday night of depriving her dying 9-year-old son medical care on religious grounds. The boy passed away from complications from diabetes in 2009. Grady showed no emotion as the verdict was read but members of her family and church were visibly upset. The jury is recommending she spend two and a half years in prison. Sentencing is scheduled for June 8, 2012.

Friday, June 8th, 2012, on *Newson6*:
A judge agreed with a Tulsa County jury and sentenced a Broken Arrow mother to two and a half years in prison Friday morning.

I have read the witness summaries. It really is a terrible read. Church members testify that they have been with Aaron and prayed for his recovery, while telling how they experienced his health. Someone tells us that he had difficulty breathing, someone that he had lost a lot of weight, someone tells us that the boy had to be carried to the toilet. Horrible reading and for me completely incomprehensible that no one had the ability to realize that the boy needed immediate care.

If you google "Susan Grady audio interview" you will find (at least as I write this part of the book) the audio recording of the hearing on June 5, 2009.

The death of Aaron comes as two other children in Oklahoma, whose parents also attend *the Church of the Firstborn*, have died, according to *Tulsa World.* According to the newspaper, Troy Damelio, 4, died at his family's trailer home in Lincoln County after being sick for a week in April, and Silas Benjamin Dobbs died hours after his at-home birth in Oklahoma City in December. No charges have been filed in either case.
The church has been linked to several deaths across the country. Austin Sprout, 16, of Creswell, Oregon, died after his parents refused to take him to the doctor after he fell ill. Although investigators would not disclose how the teen died, they revealed that he suffered from a *"highly treatable"* condition.
A Washington couple was accused of being criminally responsible for their teenage son's death for failing to call a doctor but were acquitted of second-degree murder charges in May 2012.
Zachery Swezey was 17 when he died at his Carlton home of a ruptured appendix in March 2009.
But rather than taking him to a doctor when he fell ill for several days, the Swezeys called church elders to pray for him and anoint him with olive oil.

This particular case, and a similar one about child abuse you read earlier in this chapter, are not about snakebite deaths. Both cases are included as examples of how entrenched it can be not to seek care from those who believe in the five signs. These are, of course, horrible examples; there are very few Pentecostals who refuse medical care for their children.

*

John David Brock, 60, of Stoney Fork, was bitten on the left arm while handling a rattlesnake at Mossy Simpson Pentecostal Church in Jenson. The bite, which is the latest fatal one I have found, was inflicted on July 26, 2015. I have found newspaper articles that published the news in July 2022, as a then current event, but that is wrong. Authorities said Brock refused medical treatment and that he went to his brother's home, where he died later that day. The local coroner pronounced Brock dead. Members of the church declined to comment. I have found a list from 2023 of Pentecostal churches with snake handling, the church is on the list. It is unclear if it is still ongoing.

You have been able to read about some deaths that have echoed outside the Pentecostal movement earlier in the book. I have not covered these again in this chapter, except for the latest one with a fatal outcome that was mentioned on page 25. Let's hope this was not just the latest death bite, but also the last.

*

Contrary to the perceptions of us outside of the Pentecostal community, snake handling is not considered a test of a person's faith, nor a test of God. The serpent handlers believe that they are commanded to handle venomous snakes when they feel that they have been anointed as a demonstration of God's power.

Serpent handlers are sometimes bitten, and sometimes they die.

All serpent handlers tell stories of someone who was bitten but did not suffer any ill effects from the bite. But it is really not that mysterious. Estimates are that approximately 18-22% of venomous snake bites in North America are dry bites. [16]

This not only explains why the snake handers belief in the protection of anointing appears to be supported by experience, but also explains why some don't seek medical care. Although the focus from the outside world is on snake handling, members of serpent handling churches may also drink poison (often strychnine and various pesticides) and apply fire or blow torches to their hands and faces. Deaths from poison drinking are unusual, but it happens.

Courts determine that snakebite deaths fall within four categories: suicide, homicide, voluntary manslaughter, and accidental. 33 cases of 105 total gathered, yielded information regarding the manner of death. 91% were considered to have been accidental deaths, 6% was determined to have been voluntary manslaughter and 3% were ruled to be suicide. Most of the cases were determined to be accidental but some of these cases could have been falsely reported or determined to have been accidental due to lack of evidence to prove otherwise. [17]

If you look at the number of deaths in this part of the Pentecostal movement due to venomous snake bites, the curve rises steeply during the 1920s and 1930s when the practice of using venomous snakes in churches gains ground. After that, it remains at a steady level for 40 years. The level does not fall until the 1980s, before increasing in the 1990s and remaining slightly higher well into the current century. Now the level is down to the same low level it was in the first tentative years over 110 years ago. [18]
One might wonder how large the dark figure is, especially if we are talking about old events. One might also ask whether the increase seen around the turn of the millennium could be due to the increased attention from the media, more people perhaps found faith. Finally, one can perhaps guess that the decrease we have now is due to both a reduced number of practitioners and the fact that more and more people seek care after a bite.

Since I am not a believer and I do not hunt for likes by showing pictures online of me free handling, I will continue to insist on using my snake hooks. I believe that the other method can be life shortening and my work with this book has strengthened my belief in this.
Too many people die every year from venomous snake bites, and every single death is of course a disaster. As a reader, you should not forget that. We need to remember this, even if we think that picking up a rattlesnake in church is tempting fate, and even if we think that it is madmen who die from religious handling and that they have themselves to blame. It is still the case that it is a human being who has died. A person who is mourned, a person who is missed by family and friends.

"-Do you mourn for somebody who has been serpent bit and died?"
"-Yes honey, I...the loss of them is all. Now today it's just as fresh in my memory of losing my daughter. That's something you never get used to. You just have to learn to live with it. It hurts me a lot. I shed a lot of tears about her. I know too she's safe. I feel that she's safe. It's not that I'm worried about her soul or anything, i feel like that she's really...really saved. I feel like she died right in the Lord. But still, that don't keep me from missing her."
Barbara Elkins answers a reporter's question about her daughter's death from a venomous worm bite. (I have slightly shortened the answer.) Robert and Barbara Elkins, along with her son Dewey Chapin, were the leaders of *The Church of The Lord Jesus* in Jolo, West Virginia. More on Dewey Chapin on the following pages. The daughter who died was Columbia Gay Hagerman, which you read about on pages 177-178.

I have taken part in more than one interview where the question was asked regarding the view of seeking care after a prayer, grieving family members have been asked if they wished that the deceased had chosen to seek care. Some answered that God's will be done, but some wished that the deceased family member had chosen to seek care.
The same people who had wished that a family member had chosen to seek medical care said that if they had received a life-threatening venomous snake bite, they would have chosen to put their fate in God's hands.

In the 110-plus years since George Went Hensley first picked up a rattlesnake, many have sought closeness to God through an extreme closeness to venomous snakes.
Some met their maker a little earlier than others.

May they all rest in peace.

We should remember that not everyone who handles venomous snakes in church gets bitten. Of course, most people survive, but it does not have the same impact on the media. The fact that someone wasn't bitten doesn't inspire a journalist to write an article.
My belief is that it is mostly a matter of luck and quite often a matter of snakes not being in such good condition, if one manages to frequently handle venomous snakes without being bitten. Since I don't believe in God, I also can't believe that it is God's will that someone can handle venomous snakes without consequences.

Dewey Chafin, considered by many to be the best-known snake-handler in the United States during the 1990s, is an example of how it is possible to survive a lifetime of handling venomous snakes as part of your religious practice. He starred in the 1977 documentary *"The Jolo Serpent Handlers"* and spend a lifetime handling venomous snakes. (Jolo is a small town in West Virginia.)

In an interview in 1995, Dewey Chafin said that he had been handling venomous snakes for almost 40 years. According to him, at the time of the interview, he had been bitten 118 times, but never received medical treatment after a bite. Without knowing, I would guess that he exaggerated a little, since I believe that no one can be bitten by a venomous snake 118 times and survive 118 times without medical treatment. It's also the case that a decent percentage of the bites were probably delivered by Timber rattlesnakes.

Chafin had a simple explanation to why he survived all these bites:
"I always figure God will take care of me. If he wants you to die, you'll die. If he wants you to live, you'll live. It's always the same. It's a victory over the devil. The serpent is a symbol of the devil. When you have power over it, that's God having power over evil."

Dewey Chafin claimed that he remembered nearly every bite. The first was a copperhead, which bit him on the hand in 1960.
"It hurt something terrible," he said.
He told the journalist that the most painful bite came from a black rattlesnake* that bit him on the left thumb one Sunday night in the late 1970s. The pain kept him awake for 14 days. And the most dangerous, he said, was the 5-foot-long diamondback rattlesnake (unclear whether it was a *Crotalus adamanteus* or a *Crotalus atrox*) that sunk its fangs into the skin over his right eye. *"I was pretty worried there for a while,"* he said. He received one of the bites during the filming of the documentary. The morning after the bite, he was interviewed in bed, about how he felt the morning after the bite, his thoughts on why he had been bitten once again and whether he was afraid of dying.
"-So, when you get bitten then, is it the devils work, or Gods will?"
"-Well, its Gods will but the devils work."
Dewey Chafin was convinced that when the time came, he would succumb to a snake's bite. Dewey Chafin passed away September 3, 2015, ate the age of 82. It was not a snake that ended his life.

*Probably a dark *Crotalus horridus* but could also be a *Crotalus cerberus.* Many parishioners call the darker Timber rattlesnakes "black rattlers" and the lighter ones "yellow rattlers". *Crotalus cerberus* has the common names Arizona black rattlesnake and Black rattlesnake (there are probably more). *Crotalus cerberus* has been found in the church world but is not very common.

A good ending to this part of the book could be to state that although there is an overrepresentation of venomous snake bites and related deaths, it is certainly the case that many practitioners are not bitten and that most of those who are bitten survive thanks to good luck, dry bites or medical care. One thing I'm wondering more and more is whether the phenomenon of snake handling (and associated deaths) could have happened elsewhere in the US. I think it needs to be in rural areas where it happens, with residents who like to keep to themselves and don't have too much faith in outside influences.
It probably needs to be a closed community, with its own rules and way of life.

12. Strychnine

Photo: qimono, Pixabay.
Image cropped.

Strychnine is mentioned in many places in the book, and I thought I would include a few lines about the poison. Deaths do occur after parishioners or pastors drink the poison, but it is unclear how common this is. My research tells me that it is very seldom, but deaths do occur. Strychnine is a strong poison, but it is simple to calculate what you ingest. In comparison snakes provide a somewhat unknown quantity of venom during a bite.

Mark 16:17-18, the King James version:
And these signs shall follow them that believe; In my name shall they cast out devils; they shall speak with new tongues; They shall take up serpents; and if they drink any deadly thing, it shall not hurt them; they shall lay hands on the sick, and they shall recover.

You probably know these verses by heart now that you are this far into the book. Drinking poison is mentioned in the same verses as handling venomous snakes and it is one of the five signs that are of great importance to the believers that the book is about.

"The Pastor and Congregation are not Responsible for anyone that handles the serpent's and get's bit. If you get bit the church will stand by you and pray with you. And the Same goes with drinking the poison."
Text on a piece of paper taped to the pulpit in the church where Mack Wolford preached.

The bottle, or jar, of strychnine often stands on the pulpit waiting for someone to drink from it. As with the snake handling, it is completely voluntary to drink from it and most often it is the preacher who drinks from it. The liquid is diluted and ready, often the pastor is responsible for preparing this.

I asked the always-helpful Cody Coots about the strychnine drinking and, as usual, he responded almost immediately.
"Well, if you drink it and God is not moving on you, your legs get stiff and everything locks up. One time I put just a drop on my tongue to test it and I ended up hallucinating and it messed my heart up and my heart is still messed up from that incident. If God's moving on you, it just tastes bad but no effects.
And as for the mixing it up a lot of times we what you call diluting is what we caught cutting it down. It's just because it's hard to get a hold of and we try to make as many jars as we can."

Mack Wolford talked about drinking strychnine in an interview in Washington Post Sunday magazine in 2011.
"In my life I've probably drunk two gallons of it. Once you drink it, there is no turning back. All your muscles contract at once. Your body starts stiffening out. Your lungs, it's like you can't breathe.
I was up all night struggling to breathe and move my muscles and repeating Bible verses that say you can drink any deadly thing, and it won't hurt you.
The devil said, 'You're going to die, you're going to die'.
You can't go to the hospital. There is not a lot they can do. But (seeking medical help) *means you're already starting to lose faith."*

The church members believe that there is a constant battle between good and evil, and by taking poison, speaking in tongues, and handling snakes, they are fighting the devil through the five signs. The practitioners are well aware that what they are doing could be lethal to them.

Strychnine was first used in Germany in the 16th century as a poison for rodents. It is a highly toxic, colorless, bitter, crystalline alkaloid. Strychnine is still used today in our pesticides for killing small vertebrates such as birds and rodents. Strychnine, when inhaled, swallowed, or absorbed through the eyes or mouth, causes poisoning which results in muscular convulsions and eventually death through suffocation. It produces some of the most dramatic and painful symptoms of any known toxic reaction.
The probable lethal oral dose in humans is 1.5 to 2 mg/kg [19].

Ten to twenty minutes after exposure, the body's muscles begin to spasm, starting with the head and neck in the form of *trismus* (a condition of restricted opening of the mouth) and *risus sardonicus* (a sustained spasm of the facial muscles that appears to produce grinning). The spasms then spread to every muscle in the body, with nearly continuous convulsions, and get worse at the slightest stimulus. If the mixture was too strong, or the subject drank too much of it, he/she usually dies within 2-3hrs of ingestion.

There is no antidote for strychnine poisoning. Strychnine poisoning demands (at least) early control of muscle spasms, intubation for loss of airway control, toxin removal, intravenous hydration and potentially active cooling efforts. Treatment involves oral administration of activated charcoal, which adsorbs strychnine within the digestive tract. Unabsorbed strychnine is removed from the stomach by gastric lavage. The use of activated charcoal is considered dangerous in patients with tenuous airways or altered mental states.

Strychnine is poorly soluble in water and gives the water a strongly bitter taste, which is still apparent when diluted to 1:700 000. The substance is extremely bitter, one of the most bitter substances known.

13. Afterword and thanks

Thank you for choosing to read my book! Hope you found both the subject and the snakes as fascinating as I do! It has been interesting to research and learn more about the subject of the book and I hope I have been able to pass on my fascination to you. I live in Sweden and first encountered the phenomenon of Pentecostals and rattlesnakes in the 90s, when the general public got ready access to the internet. These pastors have fascinated me ever since and getting to interview a couple of them are some of the most exciting things I've done in a long time! The believers claim that the outside world gives snakes a bigger role than they have within the church. I believe in the claim and unfortunately, this book continues to focus on the snakes more so than the church and bible.

After proofreading my text, I realize that my opinion shines through in several places, that I think it is an unnecessary risk to handle venomous snakes in church and that the snakes would have been better off not participating. I hope that I have not colored the reader's opinion too much; my idea with the book is that everyone should take a stand for themselves. I also hope that the book really was an opportunity for believers to have their say and tell their own story.

It seems to me a rather incredible idea, to risk life and limb to prove to oneself, to one's congregation or to God that one has a pure faith. Being a literalist can certainly take unusual forms. I find it hard to understand what drives people to do this, but I also find it hard to understand other forms of religious practice that limit life or jeopardize life. Just speaking in tongues, dancing as if in a trance or the like is a bit alien to me as a non-believer. I think you have to be born into the environment or be some kind of "soul searcher" to end up in what many consider religious fanaticism.
In my view, it is a practice of faith that risks jeopardizing the health of animals and sometimes even humans.

One should always be careful with generalizations and prejudices. Most of the people from Appalachia are ordinary people and most of the parishioners of those Pentecostal churches with venomous snakes are also ordinary people. They just have an unusual way of practicing their faith. Colorful and quirky people can be found everywhere, often they are the ones who are seen and noticed. It can sometimes be easy to forget that the whole world is full of pretty ordinary people. Maybe some have just grown up in a completely different environment than you and me? What is odd, strange or exotic to me is everyday life to someone else.

I hope that none of the contributors to this book have taken offence at the result. If so, I sincerely apologize, it was not my intention to step on anyone's toes. I wanted to tell, as neutral as I could, a story about an unusual phenomenon on the outskirts of herpetology.

Part of the book is history, much of it is about people from the past and snakebites from the past. To conclude, I would like to take a guess at the future of this phenomenon. The number of practitioners has been steadily shrinking and I think this will continue. Authorities and animal rights organizations will probably put an end to the phenomenon sooner or later, and if they don't, the churches that focus on the five signs will probably cease to exist with the help of membership losses. These are small churches on the countryside, and I think the US will make the same journey as Western Europe have done, fewer and fewer people go to church. In 10-15 years, there may be no more venomous snakes in the churches of the southeastern US. Maybe that is the best thing for the snakes and also for the churchgoers?

One might ask whether the fact that snake handling on the outskirts of the Pentecostal movement went viral in the early 2010s has prolonged the life of this unusual religious practice? Could it be that reality TV, the media, news articles and books are feeding this practice? In any case, since the 1980s (at least), the world has given a lot of attention to an unusual religious practice practiced by quite a few people, and this attention reached its peak just over ten years ago. I don't mean that practitioners of the faith are dealing with venomous snakes in the hope of being featured in the newspaper or otherwise getting attention outside the church hall, what I mean is that all publicity is good publicity. The attention of the outside world can help more people find this way of practicing their faith. If that is the case, then of course I contributed to this with my books.

Photo: Steve Ludwin. A congregation member holding a Timber rattlesnake, *Crotalus horridus.*

Photo: Friends and family to Cody Coots. Published with permission from Cody. Cody Coots drinks poison, unknown what he drinks but the most common in this context is strychnine.

The world's interest in snake-handling Pentecostals has faded in recent years and the number of practitioners has decreased. I wanted to capture the pastors' own stories before it was too late. If the practice continues in the future, I believe it will be on an even smaller scale than it is now, and I also believe that legislation has forced practitioners to go completely underground.

I would have liked to include more people's stories, but it is already the case that many avoid the attention of the outside world. Some of those who declined to participate are mentioned in the book, but those who also previously have avoided attention are not easy to find and are even harder to write about. Once again, many thanks to Cody and Verlin for contributing your stories. I hope you enjoy my book, despite the fact that on many points we have rather different views.

All the parishes mentioned in this book are independent parishes; no parish is responsible for what happens in any other parish. The same applies to the members of the congregation, they are not responsible for the actions of anyone else, they are not responsible for what anyone else of the same faith does or does not do.

Dr. Ralph Hood, author and professor of psychology and religion at the University of Tennessee, has followed Appalachia's snake handling Pentecostals for over 25 years. He said the following in an interview a few years ago:
"Appalachians don't like people to think that Appalachia is full of serpent handlers. Right, they don't want to be defined by serpent handling. But what they do do, that a lot of other places do, is they respect diversity and respect the handlers. None of them would be opposed to the serpent handling church being here, but none of them would support the practice."

A bit like the spirit of one of my favorite songs, *"Live and let die".* The basic meaning of the song is, you do you, and I'll do me. Whatever "you do you" entails, I'm going to let you get on with it. Don't concern yourself with other people's life choices, just be happy with your own life choices. A pretty good philosophy in many contexts in my opinion, we too often have negative views about what others do or do not do. What I find a little troubling about snake handling during church services is my belief that the snakes are sometimes harmed by what is going on.

As I wrote early in the book, my opinion is that these are not crazy people, but I think that what they do is a bit crazy. For me, there is a big difference between being crazy and doing something that is a bit crazy in the eyes of others. These believers are humans, just like you and me. The world is not black and white, people have both good and bad sides. Sometimes people make mistakes, or act in a way that others think is wrong. Heredity, environment and tradition influence how we live our lives. I find it hard to believe that something like snake handling in the Pentecostal movement could have emerged and gained a foothold today, without already resting on heritage, environment or tradition.

Religion has lost ground and animal rights have gained ground. What speaks against my belief (about the future of this snake handling phenomenon) is that in many places the world is going backwards. We are electing leaders similar to those who led countries during the interwar period, the view of women's rights has in some places been set back at least 50 years. In troubled times, many turn to religion for comfort and security. Will they also turn to snake handling?

Briefly about me

I have been interested in snakes all my life. When I was a child, it happened that I caught a Grass snake, *Natrix natrix*, or a Slow worm, *Anguis fragilis,* and took it home and had it in the garden for a while before it was allowed to move out into the forest again. In Sweden, you are not allowed to catch reptiles and take them home. I know this today, but that knowledge was lacking when I was growing up.

The species that really got me hooked on snakes was Sweden's only venomous snake, the Adder, *Vipera berus*. Fantastically beautiful snakes and at the same time a little exciting that they were dangerous!

Photo: Rickard Ljunggren. Tubing a *Trimeresurus venustus* in my snake room.

Getting a snake as a pet was not an option for me as long as I lived at home with my parents. The first reptile became a Red-eared slider *Trachemys scripta elegans* because turtles were something that the rest of my family was not afraid of. When I moved away from home, I got my first snake and since then I have probably owned over 200 snakes. Today I have just over 30 snakes in the basement, both species you can handle and venomous species. I enjoy spending time in Sweden's nature photographing vipers, and I like to search for snakes in other countries. Hiking in the nature of Southeast Asia is a pleasure I unfortunately do too rarely!
This is my ninth book about snakes and the second one written in English. I will continue to write as I find it very rewarding. You learn a lot during the hunt for facts and you build a network of contacts along the way. My next book will be about the species that made me discover this hobby and see how beautiful snakes are! It will be a book about *Vipera berus*, the only venomous snake we have in Sweden, and the genus *Vipera*. In the picture you can see a melanistic male that I encountered a kilometer from home. Although I have some specimens in terrariums in my basement, I love looking for European adders in the wild! Work has started and the aim is to complete the book sometime in 2025.
Once again, thank you for reading my book!

Rickard, January 31st, 2025

Photos next page:

Upper photo: Suzanne Petersson.
Me with a Broad-banded copperhead, *Agkistrodon laticinctus*, an adult female. It is a snake that does not occur in the wild in the Appalachian region, but the species is sometimes being used under service. Since I am a bit more careful than others who have been given space in this book, I like to use snake hooks.

Lower photo: Rickard Ljunggren.
A male melanistic Adder, *Vipera berus,* sunbathing in early spring. The picture was taken near my home, I have the privilege of living near some hibernation sites for the species.

"I've been handling snakes for about forty years. I've been bitten 151 times and still counting, I hope, probably some more will bite me. I don't get scared. I mean, sometimes one will hiss at you and you're gonna get bit, but it don't scare me none. I have handled 20 at one time. That was a long time ago... They just kept piling them in my hands, piling them in my hands. I thank God for every time I handle them, and I got them in my hands, and they don't bite me. And if they do bite me, I'm thinking, thank you God for taking care of me."

Dewey Chafin, Jolo, West Virgina, 2004.

*

I think we will let Dewey's statement from 2004 end this book. Maybe that's how it is? The confirmation of one's pure and strong faith, which one receives after surviving a bite, is as close to God as one can get? In any case, it must surely be the case that surviving a venomous snake bite strengthens one's faith because the incident has confirmed the lines in Mark's Gospel that one has placed so much emphasis on. And of course, such an incident has a great impact on other members of the congregation. Perhaps even greater because the incident took place in partially closed communities where many members come from the same families or at least are often related to each other. Could it even be seen as some kind of reward? Perhaps God has confirmed your pure faith, by allowing you to survive a venomous snake bite? At the same time, it's probably also an confirmation and/or reward every time God lets you handle a deadly snake without getting bitten. And this too confirms both to the handler and to the others that the faith is the right and the pure one.

14. References and licenses:

[1] Page 11. Anaphylactic reaction from repeated exposure to snake venom: A case report Mahmood Sasa1,2*, Adriana Alfaro-Chinchilla1, Luis Zúñiga3, Javier Valenzuela3, Raquel Martinez4. Sasa et al., Arch Clin Toxicol. Archives of Clinical Toxicology 2022;4(1):13-16.
[2] Page 18. National Institute for Occupational Safety and Health (NIOSH)
[3] Page 18. Greene, Spencer. "The Seriousness of a Copperhead Envenomation".
[4]. Page 27. National Public Radio, October 2013.
[5]. Page 27. The Jolo Serpent Handlers, 1977.
[6]. Page 64. The Guinness book of animal facts and feats by Wood, Gerald L. Page 111, The heaviest venomous snake.
[7]. Page 64. WCH Clinical Toxinology Resources, toxinology.com. Retrieved 19 March 2022.
[8] Page 67. National Institute for Occupational Safety and Health (NIOSH)
[9] Page 68. University of Georgia's Savannah River Ecology Laboratory. Green Watersnake (*Nerodia floridana*).
[10] Page 73. Handbook of Snakes of the United States and Canada by Albert Hazen Wright and Anna Allen Wright, 1957.
[11] Page 147. Biology, status, and management of the Timber rattlesnake (Crotalus horridus) : a guide for conservation. William Sandwith Brown, 1993.
[12] Page 149. Hidden life of the Timber rattler. William Sandwith Brown. National geographic 172 (1), 128-138, 1987.
[13] Page 149. Biology and status of Timber rattlesnake (Crotalus horridus) populations in Pennsylvania. John H. Galligan, W. Dunson, 1979.
[14] Page 159. Bites by Copperheads (Ancistrodon contortrix) in the United States, Henry M. Parrish, MD, DrPH; Carole A. Carr, RN, MSPH
Greene SC, Folt J, Wyatt K, and Brandehoff NP. Epidemiology of fatal snakebites in the United States 1989-2018. Am J Emerg Med 2021; 45:309-31
[15] Page 167. Shannon K. Hoss, Mark J. Garcia, Ryan L. Earley, Rulon W. Clark. Fine-scale hormonal patterns associated with birth and maternal care in the cottonmouth (Agkistrodon piscivorus), a North American pitviper snake, General and Comparative Endocrinology, Volume 208, 2014, Pages 85-93.
[16] Page 187. Russell 1983:287.
[17] Page 188. Social-Envenomation: A Ritual of Snake-Handling Churches. A THESIS SUBMITTED TO THE GRADUATE FACULTY In partial fulfillment of the requirements For the degree of MASTER OF SCIENCE IN FORENSIC SCIENCE Amy Waters Edmond, Oklahoma January 2018.
[18] Page 188. Social-Envenomation: A Ritual of Snake-Handling Churches. A THESIS SUBMITTED TO THE GRADUATE FACULTY In partial fulfillment of the requirements For the degree of MASTER OF SCIENCE IN FORENSIC SCIENCE Amy Waters Edmond, Oklahoma January 2018.
[19] Page 194. E., Gosselin, Robert (1984). Clinical toxicology of commercial products. Williams & Wilkins. ISBN 0-683-03632-7.

Page 13

I, the copyright holder of this work, hereby publish it under the following license:

This file is made available under the Creative Commons CC0 1.0 Universal Public Domain Dedication.

The person who associated a work with this deed has dedicated the work to the public domain by waiving all of their rights to the work worldwide under copyright law, including all related and neighboring rights, to the extent allowed by law. You can copy, modify, distribute and perform the work, even for commercial purposes, all without asking permission.

Page 35

This file is made available under the Creative Commons CC0 1.0 Universal Public Domain Dedication.

The person who associated a work with this deed has dedicated the work to the public domain by waiving all of their rights to the work worldwide under copyright law, including all related and neighboring rights, to the extent allowed by law. You can copy, modify, distribute and perform the work, even for commercial purposes, all without asking permission.

Page 40

Licensing [edit]

*This advertisement (or image from an advertisement) is in the **public domain** because it was published in a collective work (such as a periodical issue) in the United States between 1929 and 1977 and **without a copyright notice specific to the advertisement**. Unless its author has been dead for several years, it is **copyrighted** in jurisdictions that do not apply the rule of the shorter term for US works, such as Canada (50 p.m.a.), Mainland China (50 p.m.a., not Hong Kong or Macao), Germany (70 p.m.a.), Mexico (100 p.m.a.), Switzerland (70 p.m.a.), and other countries with individual treaties. See this page for further explanation.*

Page 43

Licensing [edit]

This work is in the **public domain** in its country of origin and other countries and areas where the copyright term is the author's **life plus 100 years or fewer**.

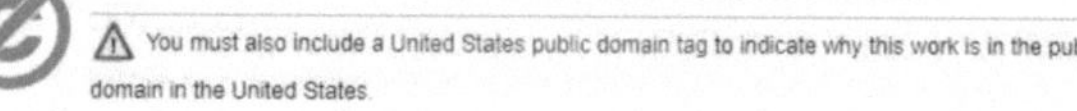

You must also include a United States public domain tag to indicate why this work is in the public domain in the United States.

This file has been identified as being free of known restrictions under copyright law, including all related and neighboring rights.

Page 44

Licensing [edit]

*This image or media file contains material based on a work of a United States Department of the Interior employee, created as part of that person's official duties. As a work of the U.S. federal government, such work is in the **public domain** in the United States. See the Department of the Interior copyright policy for more information.*

Page 45

This work is in the **public domain** in the United States because it is a work prepared by an officer or employee of the United States Government as part of that person's official duties under the terms of *Title 17, Chapter 1, Section 105 of the US Code*. **Note: This only applies to original works of the Federal Government and not to the work of any individual U.S. state, territory, commonwealth, county, municipality, or any other subdivision. This template also does not apply to postage stamp designs published by the United States Postal Service since 1978. (See § 313.6(C)(1) of Compendium of U.S. Copyright Office Practices). It also does not apply to certain US coins; see The US Mint Terms of Use.**

This file has been identified as being free of known restrictions under copyright law, including all related and neighboring rights.

Page 50

This work is in the **public domain** in the United States because it is a work prepared by an officer or employee of the United States Government as part of that person's official duties under the terms of *Title 17, Chapter 1, Section 105 of the US Code*. **Note: This only applies to original works of the Federal Government and not to the work of any individual U.S. state, territory, commonwealth, county, municipality, or any other subdivision. This template also does not apply to postage stamp designs published by the United States Postal Service since 1978. (See § 313.6(C)(1) of Compendium of U.S. Copyright Office Practices). It also does not apply to certain US coins; see The US Mint Terms of Use.**

This file has been identified as being free of known restrictions under copyright law, including all related and neighboring rights.

Page 51

This work is in the **public domain** in the United States because it is a work prepared by an officer or employee of the United States Government as part of that person's official duties under the terms of *Title 17, Chapter 1, Section 105 of the US Code*. **Note: This only applies to original works of the Federal Government and not to the work of any individual U.S. state, territory, commonwealth, county, municipality, or any other subdivision. This template also does not apply to postage stamp designs published by the United States Postal Service since 1978. (See § 313.6(C)(1) of Compendium of U.S. Copyright Office Practices). It also does not apply to certain US coins; see The US Mint Terms of Use.**

This file has been identified as being free of known restrictions under copyright law, including all related and neighboring rights.

Page 53

This work is in the **public domain** in the United States because it is a work prepared by an officer or employee of the United States Government as part of that person's official duties under the terms of *Title 17, Chapter 1, Section 105 of the US Code*. **Note: This only applies to original works of the Federal Government and not to the work of any individual U.S. state, territory, commonwealth, county, municipality, or any other subdivision. This template also does not apply to postage stamp designs published by the United States Postal Service since 1978. (See § 313.6(C)(1) of Compendium of U.S. Copyright Office Practices). It also does not apply to certain US coins; see The US Mint Terms of Use.**

This file has been identified as being free of known restrictions under copyright law, including all related and neighboring rights.

Page 55

This work is in the **public domain** in the United States because it is a work prepared by an officer or employee of the United States Government as part of that person's official duties under the terms of *Title 17, Chapter 1, Section 105 of the US Code*. **Note: This only applies to original works of the Federal Government and not to the work of any individual U.S. state, territory, commonwealth, county, municipality, or any other subdivision. This template also does not apply to postage stamp designs published by the United States Postal Service since 1978. (See § 313.6(C)(1) of Compendium of U.S. Copyright Office Practices). It also does not apply to certain US coins; see The US Mint Terms of Use.**

This file has been identified as being free of known restrictions under copyright law, including all related and neighboring rights.

This work is in the **public domain** in the United States because it is a work prepared by an officer or employee of the United States Government as part of that person's official duties under the terms of *Title 17, Chapter 1, Section 105 of the US Code*. **Note: This only applies to original works of the Federal Government and not to the work of any individual U.S. state, territory, commonwealth, county, municipality, or any other subdivision. This template also does not apply to postage stamp designs published by the United States Postal Service since 1978. (See § 313.6(C)(1) of Compendium of U.S. Copyright Office Practices). It also does not apply to certain US coins; see The US Mint Terms of Use.**

This file has been identified as being free of known restrictions under copyright law, including all related and neighboring rights.

Page 61

Licensing [edit]

This work is in the **public domain** in its country of origin and other countries and areas where the copyright term is the author's **life plus 70 years or fewer**.

This work is in the **public domain** in the United States because it was published (or registered with the U.S. Copyright Office) before January 1, 1929.

This file has been identified as being free of known restrictions under copyright law, including all related and neighboring rights.

Page 63

The author died in 1805, so this work is in the **public domain** in its country of origin and other countries and areas where the copyright term is the author's **life plus 100 years or fewer**.

This work is in the **public domain** in the United States because it was published (or registered with the U.S. Copyright Office) before January 1, 1929.

This file has been identified as being free of known restrictions under copyright law, including all related and neighboring rights.

Page 143

This work is in the **public domain** in its country of origin and other countries and areas where the copyright term is the author's **life plus 100 years or fewer**.

⚠ You must also include a United States public domain tag to indicate why this work is in the public domain in the United States.

This file has been identified as being free of known restrictions under copyright law, including all related and neighboring rights.